BUILDING JAPAN : 1868–1876

RICHARD HENRY BRUNTON

BUILDING JAPAN 1868-1876

WITH AN
INTRODUCTION & NOTES
BY
SIR HUGH CORTAZZI, G.C.M.G.

In addition to the 1906
INTRODUCTORY, POSTSCRIPT & NOTES
BY
WILLIAM ELLIOT GRIFFIS

LONDON AND NEW YORK

First published 1991 by Japan Library Limited

2 Park Square, Milton Park, Abingdon, Oxon OX14 4RN
711 Third Avenue, New York, NY 10017, USA

Routledge is an imprint of the Taylor & Francis Group, an informa business

First issued in paperback 2016

BUILDING JAPAN: 1868-1876

Copyright © 1991 Japan Library Ltd

ISBN 978-1-138-96518-8 (pbk)
ISBN 978-1-873410-05-9 (hbk)

British Library Cataloguing in Publication Data
A CIP catalogue record for this book
is availble from the British Library

Set in Bembo Roman 13 on 14 point
Keywork by Ann Tiltman
Photosetting by Visual Typesetting

CONTENTS

PUBLISHERS' NOTE

THE TEXT of Brunton's work printed here is as edited around 1906 by the American teacher William Elliot Griffis who, in the Postscript, points out that the text available for publication is his 'condensed and annotated' version of the original Brunton manuscript. Furthermore, Griffis admits to leaving out the complete first half of Brunton's record which he describes as 'historical in form' and presumably, given all the existing histories in print at the time, was not considered relevant, nor a commercial proposition.

It is therefore not at all clear how much rewriting Griffis has done, nor how much of the original Brunton text has been omitted, which as far as is known, no longer exists.

Griffis naturally edited the work according to American spelling and stylistic conventions — as well as perceived market opportunity, given Japan's high profile worldwide after her success in the Russo-Japanese war of 1904-5. Since Brunton would have written in British English, we have tried to honour the style of the original author and returned to British English spelling.

In the same way, preparing the book for an American audience, Griffis changed all Brunton's references to costs from pounds sterling to American dollars, except on a few occasions when he failed to do so. Here we have not attempted to impose any standardisation and left the figures as given by Griffis.

On the other hand, to mark the fact that this is a 'period' work, published in its original form for the first time, the nineteenth-century spelling conventions for the names of Japanese towns and cities, e.g., Tokio for Tokyo, Kioto for Kyoto have been retained, as well as the extensive use of capital letters and the convention for expressing the date. However, the then common practice of accenting the live 'e' as in Kobé and saké has not been followed.

Footnotes

The text references to the original Griffis footnotes are printed in Bold (heavier type); the text references to the footnotes by Hugh Cortazzi are printed in Roman (lighter type).

LIST OF ILLUSTRATIONS

following page 86

PREFACE

BRITISH civil engineer Richard Henry Brunton (1841-1901) came to Yokohama in 1868, the first year of the Meiji Era, and shortly after the port of Yokohama was opened to foreign trade. He was the first of a number of so-called *o-yatoi-gaijin* (foreign employees) contracted to the Meiji government, which was seeking to quickly modernise Japan by introducing foreign technology. R. H. Brunton's work was to construct lighthouses throughout Japan, based on treaties concluded between the then defunct Tokugawa shogunate and the American and European governments. To promote foreign trade, it was necessary to build lighthouses for the safe navigation of vessels that entered and departed from the Japanese ports that had already been opened.

Brunton's efforts were outstanding. He constructed more than 30 lighthouses in important spots throughout Japan during his rather short eight-year stay until his departure from Japan in 1876. Being among the first foreign professionals to be employed by the Meiji government, he inevitably ran into many difficulties in his relations with Japanese government officials who were drawn exclusively from the class-conscious warrior 'samurai' tradition.

Brunton had been engaged in railway construction during his earlier days in Great Britain, and had established a broad civil engineering career. Thus, he was asked for technical advice on railway construction as well as telecommunications, harbour works and many other projects being undertaken by the Meiji government. In particular, because the Lighthouse Department to which Brunton was appointed was located in Yokohama, the city became centre stage for modern Western engineering transplanted to Japan by Brunton. In the central part of Yokohama City we can still see much of Brunton's legacy today in roads, parks, drainage systems and bridges, and we are proud to recognise them as 'Japan's firsts'.

This year marks the 150th anniversary of Brunton's birth. On this occasion, in recognition of his great contributions that made Yokohama a modern city, we decided to organise various Brunton anniversary events in cooperation with the Japan Society of Civil Engineers, the Yokohama Chamber of Commerce and Industry and many other interested parties. Our plan includes a series of programmes both in Japan and Great Britain highlighting Brunton's achievements.

In his later years Brunton, known as the 'Father of Modern Engineering in Japan', wrote his memoirs about his experiences in Japan. We feel it most meaningful and an honour for us to be able to assist in the publication of this valuable record in England this year, as part of our Brunton memorial events.

In closing, I must state my deep gratitude for the courtesy and support of former British Ambassador to Japan, Sir Hugh Cortazzi, and many others involved in this project.

HIDENOBU TAKAHIDE, DR. ENG.
Mayor
City of Yokohama
SUMMER 1991

INTRODUCTION

BY HUGH CORTAZZI

RICHARD Henry Brunton was born in Fetteresso, Kincardineshire, Scotland, in December 1841. His father was a Captain in the Royal Navy.

After attending private schools in Scotland, Brunton in 1856 began his articles as an engineering student under a Mr P. D. Brown in Aberdeen; in 1858 he was transferred to the supervision of Mr John Willet also of Aberdeen. He completed his articles in 1860 and remained in Mr Willet's employ working on the construction of various railways in the Scottish Highlands. In 1864 he moved to London and was employed first on parliamentary work in connection with the London and South Western Railway. In 1866 he became principal assistant to a Mr Henry Bolden and was involved in work for the Midland and other railways.

In February 1868 Brunton, who became an Associate of the Institution of Civil Engineers in April that year, was appointed, on the recommendation of Messrs David and Thomas Stevenson of the Scottish Lighthouse Board, as Chief Engineer to the Lighthouse Department of the Japanese Government. He arrived in Japan in August 1868 and except for a brief period of home leave remained there until 1876. While in Japan his main responsibility was for the construction of lighthouses but he also became involved with a wide variety of civil engineering projects.

On his return to Britain, Brunton was appointed in 1878 Manager of Young's Paraffin Oil Company in Glasgow. In 1881 he bought a manufacturing business producing architectural ornaments and practised in

London as an architect and engineer. He died in London on 24 April 1901.

To appreciate Brunton's achievements in Japan it is necessary to understand something of the background.

When the ships of the US Navy under the command of Commodore Perry appeared off the coasts of Japan in 1853 the country had been virtually closed from the rest of the world for over two centuries. The shogunate or *bakufu*, the military feudal government of Japan, in Edo (the modern Tokyo) was under the nominal leadership of the Tokugawa family who provided the shogun. The regime had become ossified and was incapable of initiating radical changes or responding positively to the potential threat of foreign domination.

The first treaty of 'Peace and Amity' with the United States, obtained by Commodore Perry in 1854, which was followed by a similar treaty with Britain, did little more than open a small window for contacts with the outside world. The foreign powers were not satisfied and in 1858 Townsend Harris, the first US Consul General in Japan, managed to conclude a commercial treaty which provided for the development of trade and the establishment of a number of treaty ports where foreign merchants could reside. Lord Elgin from Britain followed up on the American success and concluded a similar treaty between Britain and Japan.

By the treaties of 1858, the Japanese were forced to grant extra-territorial rights to foreign governments who thus had jurisdiction over their own nationals in the treaty ports. The most important of these treaty ports in the 1860s was Yokohama although there were also small settlements in Nagasaki (Kyushu) and Hakodate (Hokkaido). Kobe was opened for trade from 1 January 1868 and Niigata in 1869.

A supplementary convention was concluded in 1866 which stipulated that the Japanese government were to provide 'such lights, buoys, or beacons, as may be necessary to render secure the navigation of the approaches' to the treaty ports. This was the basis for Brunton's employment in Japan.

In the 1860s when the United States was preoccupied by the Civil War, Britain displaced the United States as the leading foreign power in relations with Japan. The British merchants quickly became predominant in Yokohama. The settlement in those early days suggested a Western frontier town. The Japanese authorities were unable to provide adequate security against anti-foreign elements and a British garrison had to be established. Murders and violence were frequent especially in the first half of the 1860s. Housing and sanitation were at best primitive. The merchants were tough and often unscrupulous. Relations between the British Minister at the Legation which was located for part of the time in Edo and the community in Yokohama were bad.

The Tokugawa shogunate's power over Japan's feudal magnates was undermined by their inability to cope with the foreigners. Their vacillations and prevarications incensed the foreign envoys and infuriated the anti-foreign elements which were predominant especially in the so-called outer domains (those of the *tozama daimyo* who were not direct vassals of the Tokugawa). The two most important of these were Satsuma (the modern Kagoshima in southern Kyushu) and Choshu (now part of Yamaguchi prefecture).

The murder in 1862 by samurai from Satsuma of a British merchant from Shanghai, Charles Richardson, at Namamugi near Yokohama on the Tokaido, the main route from Edo (Tokyo) to Kyoto, led to British demands for reparation and punishment of the murderers. When Satsuma refused to comply with these demands and the *bakufu* were unable to enforce their authority a British fleet bombarded Kagoshima. Satsuma claimed a moral victory when the fleet suf-

5

fered losses and withdrew, but the young samurai leaders in Satsuma recognised the realities of British power and enmity was translated into friendship while Satsuma strengthened its forces for the imminent struggle with the Tokugawa. It took somewhat longer for Choshu to recognise where their best interests lay. In 1864 the straits of Shimonoseki having been closed to foreign shipping by the Choshu authorities, a Western naval force bombarded the Choshu batteries which were later destroyed by a British landing party.

By 1867 the anti-Tokugawa forces had gained the upper hand and the last of the Tokugawa shoguns agreed to resign, but the shogun's allies refused to accept what they saw as a humiliation at the hands of the outer fiefs and civil war ensued in 1868. The Tokugawa forces in Edo gave in and by the end of 1868 the civil war was virtually over although resistance continued into 1869 in Hokkaido. The new regime in which samurai from Satsuma and Choshu were dominant restored nominal power to the Emperor and moved the imperial capital from Kyoto to Edo which was renamed Tokyo. This marked the inauguration, in early 1868, of the so-called Meiji [meaning enlightened government] Restoration which was a major turning point in Japanese history.

The young samurai from the outer fiefs who dominated the new regime were determined to revise the unequal treaties with their provisions for extraterritoriality. Japan must above all avoid the fate of China where the Western powers were grabbing concessions. They wanted to make Japan a significant power in the Far East. To achieve these objectives, Japan had to be dragged as quickly as possible from a semi-medieval form of feudalism into the world of the nineteenth century. Radical changes had to be made in forms of government. The rigid class structure of Tokugawa Japan dividing the people into samurai, farmers, artisans and merchants where commerce was a despised occupation had to be destroyed. A modern economic structure and an industrial base had to be built. Above all Japanese must be educated to compete on terms of equality with the West.

Japan's own resources were very limited. In a few of the fiefs a rudimentary start had been made in building a few industrial enterprises. Education had been almost entirely Confucian but literacy rates were comparatively high by nineteenth-century standards. Loyalty was regarded as the highest virtue. Strict discipline was maintained and Japanese were generally dedicated and hard-working. However, a revolution in attitudes and the radical reforms which were necessary could not be achieved easily. Opposition to change culminated in the Satsuma rebellion of 1877.

The *bakufu* had realised that Japan needed foreign help in the modernisation process and took steps to employ some foreign experts. The new leaders intensified the process. Technical experts and teachers were sought from Europe and from the USA. These were the so-called *o-yatoi-gaijin* or foreign employees.

The British Minister to Japan from 1865–1883 was Sir Harry Parkes. Domineering and arrogant, he was a determined upholder of British interests. In the run-up to the Restoration he had leaned towards the anti-Tokugawa forces. After the Restoration he did his best to ensure that British influence was maximised by ensuring that British candidates for posts with the new regime were preferred. Partly at least as a result of his efforts the British accounted for half of all foreign nationals employed by the Meiji government. Until the Ministry was abolished in 1885, two-thirds of all foreign employees in Japan were hired by the Ministry of Public Works. In order to attract the necessary experts the Japanese had to pay comparatively high salaries (often more than double the average in the USA and Europe). Understandably, as soon as Japanese replacements could be found and trained the foreigners were replaced. So the number of foreign employees was greatest in the early years after the Restoration. A total of some 3,000 foreigners were employed by the Japanese government during this period.

Despite the high salaries paid, the job of the foreign employee was rarely an easy one. There was inevitably a serious linguistic problem. The cultural barrier presented even greater difficulties to both the foreign employees and their Japanese employers. The foreign employees objected to interference and dictation from Japanese authorities who failed to understand the nature of the problems to be tackled. The Japanese for their part resented the often condescending and arrogant attitude of some of the foreign employees who were the creatures of their age and who were not always as sensitive as they should have been to Japanese susceptibilities. Despite these difficulties the foreign experts made a very significant contribution to the modernisation of Japan.

Among the many British experts and teachers invited to Japan, Richard Henry Brunton was outstanding. His achievements in the creation of the Japanese lighthouse and lifeboat service was of great importance for the safety of life at sea and navigation in Japanese waters. This facilitated the expansion of trade with Japan and benefited foreign countries as well as Japan. His help in starting the first Japanese telegraph and with plans for the building of railways in Japan was significant for the development of communications in Japan. His proposals for better access to the ports of Osaka and Niigata and for improvements at Yokohama were not unfortunately carried out during his time in Japan but his involvement with the construction of the first iron bridge in Yokohama together with his plans for the provision of water, drainage and lighting in Yokohama make him one of the great pioneers of civil engineering in Japan.

Brunton's story is a revealing one. He encountered numerous difficulties and frustrations during his service in Japan. He was a man of principle who was not prepared to accept compromises which he felt betrayed those principles. He was a difficult man to work for and to have as a subordinate. He was very determined and obstinate. He was also intolerant and impatient and sometimes arrogant. But he was very energetic, conscientious, courageous and tough. Furthermore, he had vision and was an eminently practical civil engineer.

Brunton summarised his difficulties with other foreign employees and with his Japanese colleagues and assistants in these words: 'The conscientious and efficient conduct of work in Japan was a task which presented the most perplexing difficulties. Their [i.e. the foreign employees] high pay, their different mode of living, their want of disciplinary power, and the knowledge that foreigners were more or less indispensable to the Japanese rendered their European assistants most intractable and difficult to deal with. Resignation, insubordination, absence from duty, drunkenness, and other aberrations of conduct among Europeans employed in the Japanese government service, became frequent and distressing. On the other hand, the semi-ignorance of the native servants of the Emperor, and the self-esteem, untrustworthiness, craftiness and corruption of the Japanese underlings rendered cooperation by an honourable foreigner with them extremely irritating.'

Brunton's assessment of the Japanese at that time was shared by other observers. One of his subordinates [unpublished and unsigned manuscript, a copy of which the writer received from Dr Wallace] was patronising in a manner typical of the less attractive features of our Victorian forbears. He declared: 'taking them altogether, I suppose they are the most advanced of all half-civilised nations and they might be much more advanced if they could get quit of …their self-conceit, for that will be a very great drawback to their advancement in all civilised customs. They, as a race, are particularly well pleased with themselves. They have the Chinese idea that they are superior in most everything to the whole world and they look upon all foreigners as their inferiors.'

Brunton decided that he had two choices. 'The first choice, which promised quietude and repose, was to let things take their course, give advice when asked for, feeling undisturbed if this was neglected, and to become imbued with the Oriental estimation of the valuelessness of time, allowing nothing that hinders progress to perturb or annoy. Such was the method, in the seventies, by which the European could become a favourite with his Japanese employer.' The second choice was to insist on 'a due enforcement of his directions'. Brunton noted that this line was almost certain to create friction. He commented that 'the Japanese had made up their minds to make what use they could of their foreign servants, but in no case to have them become masters, or to invest them with any power. They hold them in the position of advisers or instructors only, without the authority to direct.'

The Japanese attitude is understandable but we can also sympathise with Brunton's determination that, 'at whatever personal discomfort or self-sacrifice, I should assert my position as the responsible conductor of operations'. He had to confess that he met with only partial success.

Shortly before he died Brunton prepared an account of his service in Japan under the title 'The Awakening of a Nation, being a description of the entry of Japan into the Sisterhood of nations, with an elucidation of the Character of the people, from personal experience'. The manuscript was bought from Brunton's widow for twenty pounds by William Elliot Griffis (1843-1928). Griffis was an American teacher, clergyman and author of many books on Japan including *The Mikado's Empire* (1876). He had been employed in Japan between 1870 and 1874 and had known Brunton. Griffis worked on the manuscript and tried without success to get it published. He gave it the title 'Pioneer Engineering in Japan: A Record of Work in Helping to Re-Lay the Foundation of Japanese Empire (1868-1876)'. On Griffis's death the manuscript with other Griffis papers was left to Rutgers University, New Brunswick, New Jersey, from which Griffis had graduated. Other scholars have worked on the manuscript since. They include a Miss Stoper, Mr Frederick Weldon and Edward R. Beauchamp. A Japanese version was published in 1986 under the title *O-Yatoi Gaijin no Mita Kindai Nihon* by Shintaro Tokuriki.

The text of Brunton's work printed here is as edited by Griffis. Griffis's notes and introduction have also been included (as far as possible). I have added some further notes to help the modern reader.

The appendices include the texts of articles and papers by Brunton which are only available to students with access to specialist libraries. The text of Brunton's lecture in 1876 to the Institution of Civil Engineers on the subject 'The Japan Lights' is also included with an abstract of the discussion upon the paper although this was republished as an appendix to *The Life and Times of The Illustrious Captain Brown, a Chronicle of the Sea and of Japan's Emergence as a World Power* by Lewis Bush in 1969 (Voyagers' Press Limited, Tokyo and the Charles E. Tuttle Company, Tokyo and Rutland, Vermont).

The spelling of place names used in the Brunton manuscript has been retained. Brunton, for example, uses the spelling Tokio for Tokyo, Kioto for Kyoto, Yedo for Edo (the modern Tokyo).

The writer and publishers wish to record their sincere thanks to Dr Hidenobu Takahide, Mayor of Yokohama, to Mr Yutaka Uyeno, Chairman of the

Yokohama Chamber of Commerce, of the Brunton Memorial Committee and of Uyeno Unyu Shokai, to all members of the Committee and to the officials of the Yokohama City Government involved in this project for their generous support and assistance which has enabled this historically important work to be published. They also with to acknowledge the support of the Japan Society, London.

Finally, we would like to acknowledge the role of Mr Hiroshi Isohata of NKK Corporation who, as a result of a personal interest in early British engineers while attached to his company's London office in the 1980s, discovered that Richard Brunton and his work in Japan had been largely forgotten in Britain. Mr Isohata's quest to 'rediscover' Brunton ultimately led to the 1991 Brunton celebrations, marking the 150th anniversary of his birth.

SELECTED BIBLIOGRAPHY

BARR, Pat: *The Coming of the Barbarians*, Macmillan, London, 1967.
BARR, Pat: *The Deer Cry Pavilion*. Macmillan, London, 1968.
BEASLEY, W. G.: *The Meiji Restoration*, Stanford University Press, California, 1972.
BEASLEY, W. G.: *The Rise of Modern Japan*. Charles Tuttle, Tokyo, 1990.
CHECKLAND, Olive: *Britain's Encounter with Meiji Japan, 1868-1912*. Macmillan, London, 1989.
CORTAZZI, Hugh: *Dr Willis in Japan, British Medical Pioneer 1862-1877*. Athlone, London, 1985.
CORTAZZI, Hugh (edited): *Mitford's Japan: The Memoirs and Recollections, 1866-1906, of Algernon Bertram Mitford, the first Lord Redesdale*. Athlone, London, 1985.
CORTAZZI, Hugh: *Victorians in Japan: In and Around the Treaty Ports*. Athlone, London, 1987.
FOX, Grace: *Britain and Japan, 1858-1883*. Clarendon Press, Oxford, 1969.
JONES, H. J.: *Live Machines: Hired Foreigners in Meiji Japan*. Paul Norbury Publications, Folkestone, co-published with the University of British Columbia Press, Vancouver, 1980.
PEDLAR, Neil: *The Imported Pioneers: Westerners who helped build modern Japan*. Japan Library Ltd., Folkestone, 1990.

BRUNTON, THE LIGHTHOUSE BUILDER

WILLIAM ELLIOT GRIFFIS

WHEN IN 1866, after a century-and-a-half of interior intellectual preparation, awakening Japan called for help, England responded nobly. One of the grandest instances in history of the Macedonian cry, 'Come over and help us', was met by a most generous British response in brain, in heart, in character, in industry. Having in the seventeenth century given Will Adams[1] the shipbuilder, Saris[2] the navigator and Cocks[3] the trader, she kept up even a nobler succession in the century of steam and electricity.

Besides a body of promising young students fresh from the universities, whose names, Mitford,[4] Satow,[5] Aston[6] and others, shine like stars in the bright sky of achievement, the Queen's Government sent its ablest servant, Sir Harry Parkes,[7] who, educated from boyhood in China, knew well the history of Japan. Penetrating beyond the impudent forefront of sham and hoary deceit in Yedo, he saw the rising sun of the coming new Japan before it emerged above the horizon. Its rays had already struck the white peaks of recovered ancient history. They were soon to illuminate the path of the future. Parkes saw and endured when to many other men there was darkness with no promise. With all his fiery soul, Sir Harry threw himself into the struggle, seized opportunity by its foremost hair, and wrestled in manful grip until victory was won. He himself told me the story of how, on arriving in Japan, in June 1865, in succession to Sir Rutherford Alcock,[8] he went at once to Satsuma[9] – that storage battery of forceful Japan – and at possible risk of life found out who was the true sovereign of Japan and where the line of the future lay. On his initiative and direct suggestion, British capitalists loaned money to finance the new ventures such as the Mint,[10] Railway[11] and other Public Works of the quickly formed Mikado's Government of 1868. For years his resistless energy urged the Japanese in Tokio onward to new horizons of vision and vaster issues of action. Especially was he strenuous in persuading them to discard oriental suspicion, keep promises, abandon pagan bigotry, insular narrowness and barbarous practices, and to carry to completion grand resolves.

It was Sir Harry Parkes, chief in foresight, vigour and originality, who, even before the Mikado's Government was formed in Kioto in 1868, secured the new agreement of 1866, which supplemented and completed that made in the Townsend Harris Treaty of 1858, that had opened the five ports to foreign trade and residence. The stipulation of 1866 called for the lighting of the coast, the paving, draining and lighting of the new settlements and the general adjustments of a modern commercial civilisation to a nation in a state of medieval hermitage. Night and day, year in and out, Sir Harry championed reform, fought the battles in council and backed ever with vigilant eye and cheering word Mr R. H. Brunton, the man who lighted the coasts of Japan. This enterprise was his own child, and so Sir Harry called it. As passionately eager and earnest as a mother, he watched and served at the cradle of the young enterprise.

Yet even under the new Government, during the first decade of its existence the habits of ages of officialdom in Japan (much like that which still curses Russia) could not be changed in a day. Dishonesty, peculation, the wanton pride of the strutting creature dressed in brief authority, disdain of inferiors, contempt of the man who was educated to work with his hands, adherence to old methods of routine, and impotent fear and often spiteful jealousy of the alien, too frequently marked the average man in office. Fed from the public crib, the underling usually owed his appointment to the instincts of clanship, political patronage, or the accidents of birth, rank, or following, rather than to character or fitness. The spirit of 'graft' was everywhere. 'Wet hands in bran' was the popular proverb applied to the greedy office-holder. I speak from four years (1870–1874) of personal experience of the time, for I served the Mikado's or Imperial Government of Japan, when it was but an infant of days.

Against the abominable and wasteful systems then in vogue, when the Japanese Ship of State was overloaded with hangers-on and office-holders who must needs be parasites on the Treasury, while the number wanting to get into office, nearly swamped the vessel, none fought more bravely than the new native leaders. These men trained in the Oyomei[12] philosophy, which sets morality first and joins action to knowledge, persevered for honesty and economy. They introduced public budgets and systematic bookkeeping, reduced the number of paid officials to about one-tenth the ancient number. They assured steady progress through economy and reform. No harder work was that of Okuma,[13] Okubo,[14] Ito[15] and Inouye,[16] than that of honest Brunton.

One of the men who helped mightily by word and voice, by pen and petition, by firm resistance to dishonest or truculent officialism and who in his life and work gave unforgotten lessons of honour, integrity and honest work was Mr Richard Henry Brunton, son of Captain Brunton R.N., known also as a writer of sea-stories, his sister being the mother of Edmond Yates, editor and novelist. The aunt of Captain Brunton was the Countess of Craven. Born in Aberdeenshire in December 1841 and educated at private schools in Scotland, R. H. Brunton had nearly ten years practical experience as an engineer in railway survey and building in various parts of Great Britain. Of a thoroughly practical mind and well trained for his future work, he was appointed lighthouse engineer for Japan, at the age of twenty-seven, in April 1868. By abilities and temperament, Mr Brunton was admirably fitted to do the work in Japan which he fulfilled so well, and the story of which he has told so clearly.

Brunton hardly saw the better side of Japanese character. His contact was with the old Japan, scarcely emerged from feudalism, in which the merchant and trader had no social position and were as dirt under the samurai's foot. Feudal ideas had ruled for seven hundred years, breeding an almost inconceivable pride in the men who paid no taxes or tolls, strutted about in silk and inspired terror with their swords, keeping down the common people as masters above serfs. Another reason for official conceit was that while the mechanic was honoured above the trader, and the farmer was higher socially than the artisan, yet anything like technical labour, apart from the making of swords or works of taste or art, was despised. Admirable carpenters and workers in the lines of stone and metal for native needs there were in abundance; but builders in the large sense, blacksmiths, plumbers, glassmakers, opticians and men especially skilled in the trades necessary to modern machinery scarcely existed. Nor could the needs of commerce and the country, shrouded at night in darkness, await the slow process of training native artisans.

It was the old world of Japan, very much like that of feudal Europe, in which, despite the long peace and an industrialism peculiar to the hermit and island people, there was little at hand to aid the European master of lights. Brunton came to build lighthouses. Pressing necessities compelled him to design, build, launch and equip ships, to make bridges, project railways, transform swampy land into a modern city, pave, drain and light new settlements, and, when cargoes of machinery were wrecked in the ocean, to assemble and set up the lamp posts of the sea for the guidance of navigators by making lighthouses out of ship's lenses and American locomotive headlights. The new Japan, that at Yokosuka[17] would build and launch, on 15 November 1906, a battleship of 19,000 tons, owes much to the man who trained some of her first modern mechanics and built her first lightships.

Brunton, when his work was done, had audience of the Emperor, a banquet and present in money, but he was never decorated. This may have been because the Bureau of Decorations, which showers its favours even on obscure petty attendants, upon political envoys and missions, had not then been formed. One wonders also whether the Japanese ever decorate their critics - to whom they owe so much. Praise and flattery undoubtedly have cheered and helped this most interesting people of Asia, but it is no less certain that intelligent and friendly criticism has stimulated them even more.

The feverish haste of the first generation of the Japanese people whose eyes were opening to the wonder world of modern machinery, inventions and harnessed force, was often as comical as distressing in its manifestations. Glee and childish delight were often suddenly changed to sorrow and disaster. Wreck, explosion, poisoning, disease and mutilation forced them to realise the might of evil in untamed energies, of the nature of which they were so profoundly ignorant. It took them long to realise the necessity of scientific training and preparation to master the dragon on whose back they would ride.

Brunton tells more than one comical incident of this sort in his fascinating narrative. On the other hand, it was not very easy for Europeans and Americans,

17

whose conceit quite equals the Asiatics', and who came with preconceived ideas about 'Orientals', to realise the temper of the Japanese. Few foreigners in the seventies, apart from a handful of discerning scholars, had any idea of the century or more of interior intellectual preparation[1] in Japan for meeting the needs confronting the twentieth century. Very few Occidentals to this day, despite Japan's vantage in arms on sea and land against China[18] and Russia,[19] are acquainted with *Bushido*, or know anything of the course of philosophy, as expounded and exemplified by wise and good men in Japan, which armed them to meet the problems of a readjustment of ideals in civilisation. From the very first, though clearly seeing the need of radical reform and the desertion of traditional orientalism, the Japanese leaders determined to give their foreign *yatoi*,[20] or hired foreign servants, liberal wages but not a shred of power. Most fortunately for their sound political training and happily for their real and ultimate success, but for thirty years until 1897,[21] most galling to their pride, the Japanese were not allowed in the treaties to govern foreigners resident on their soil or in their seaports. Nor did they gain full sovereignty and equality before the nations until Great Britain, under Lord Salisbury's Government, became the bold and magnanimous leader in this act of justice.

In this jealousy and vigilant guarding against any holding of power by their salaried foreign servants, were the Japanese any different from British or American patriots? Surely not! Besides, these 'Mikado-reverencers', now exultant in national unity, had had during the ages enough of usurpers, of Tycoonery[22] (1184–1868), of powerholders, and of encroachments on the Emperor's prerogative and on national sovereignty. If they were sometimes strenuous, and even, as it appeared, through over-officious underlings offensive, it must be said that they were no more severe against foreign servants than they were against their own people. Let the instant jealousy in Tokio of 'local authorities' in 1870, the abolition of feudalism in 1871, to both of which incidents I was witness, the putting down of local reaction and insurrections, and finally the suppression of the great uprising of 1877, at so awful a cost of blood and treasure, prove this. Even-handed was Japanese justice in its unquailing resolve to possess full sovereignty. Tycoonism had cursed their country long enough, nor had the behaviour or political policy of Europeans in Asia inclined the Japanese to confidence in foreigners. Even more perhaps were they anxious not only to command success in this their vast venture, but to deserve it. Nevertheless for their 'dishonourable secrecy', past or present, we make no plea. Honest Brunton was no more irritated than was honest Townsend Harris.[23] The love of truth for its own sake is still Japan's most crying need. Faithfulness to contract is still the exotic virtue which her sons in trade and commerce need to cultivate.

Brunton's own pronounced limitations appear in his own story and need not be enlarged on. He was a man of conscience who was impatient with anything but good work. He may have been a severe disciplinarian, but that was what both his employers and his workmen needed. Happily also he had noble men in high office to reinforce his reasonable demands. Sanjo,[24] Iwakura,[25] Kido,[26] Terashima,[27] Sano,[28] Okuma, Ito and others. I knew them all, and noble characters they are and were. I knew Brunton, too, during four of his years in Japan, when we were fellow-servants of the Mikado.

In a large sense, Japan is the pupil of the English-speaking nations. The line of work done and influence generated by Great Britain and the United States is long and honourable. In the full result, we see that each atom's force 'moves the light-poised universe'.

From Will Adams to Prince Arthur,[29] who girded the Japanese Emperor with the order of the Garter and hung the Mikado's silken sun flag in Windsor Castle, the line is a noble one. Will Adams, Saris, Cocks, were English. Then follow the hundreds of American whale hunters. Yet the Scotsman cannot be left out. How odd that Ranald MacDonald,[30] half Chinook Indian whose father was a Hudson's Bay Scotsman, the first in the forties to be set voluntarily adrift and ashore in Japan, should be the first teacher of English! So it was. He trained the interpreters to use the tongue of the mother country and the Great Republic. It was MacDonald's telling about and making possible Japanese knowledge of inventions such as steamships, railways and telegraphs, that kindled the curiosity of the Japanese to know and possess.

Their opportunity came in March 1854, after Commodore M. C. Perry, formerly chief officer of the United States Lighthouse Service, had made the treaty. On the strand at Yokohama, where now stand the United States Consulate and Union Church, what may be called the first Industrial Exposition in Japan was held and the presents put on exhibition. An American baby railway, with engine, tender, passenger car and rails complete and in operation, under engineers Gay and Danby carried *one* fat Japanese passenger *on the roof*. A telegraph, manufactured by Messrs Draper and Williams, the wires duly set up on poles, sent messages in English, Dutch and Japanese. Firearms, lifeboats, balances, grain measures, agricultural implements, maps, dictionaries, etc., etc. completed the exhibition and tally of gifts. Fourteen years later, Brunton came to turn toys into realities and advance the kindergarten to a university. He found Japan very much as England was when Will Adams left it, but is it too much to say that during the Meiji period (1868-1906+) Japan has progressed in material civilisation as much as has England since the days of Elizabeth?

BUILDING JAPAN 1868-1876
BY
RICHARD HENRY BRUNTON

Also titled:

'The Awakening of a Nation: being a description of the entry of Japan into the Sisterhood of Nations, with an Elucidation of the Character of the People, from personal experience'

[BRUNTON'S ORIGINAL TITLE]

'Pioneer Engineering in Japan: A Record of the Work in Helping to Re-Lay the Foundation of Japanese Empire (1868–1876)'

[TITLE GIVEN BY GRIFFIS]

Brunton in his later years
Photograph Maull & Fox, Piccadilly

CHAPTER ONE

MY APPOINTMENT TO JAPAN

ONE OF THE FIRST acts of the British Minister, Sir Harry Parkes, on his arrival in Japan in 1865, was the conclusion of a convention supplementary[1] to the then existing treaties, which made provision for various requirements necessary to the safety and well-being of Europeans and Americans. This stipulated that the Japanese Government 'shall provide all the ports open to foreign trade with such lights, buoys or beacons, as may be necessary to render secure the navigation of the approaches to said Ports'. He pressed the ministers in Yedo to give the latter clause their attention, at the earliest opportunity. By diligent enquiry, he obtained from the captains of the warships of various nationalities their opinions as to the best position for the buoys, lights and lighthouses. On the 17th of November, 1866, he sent these to the Yedo Government, with an explanatory dispatch.

To this the Shogun's ministers replied on the 7th of December, 1866. 'It is impossible to decide where the lighthouses should be erected until accurate enquiries shall have been made, but, in the meantime we intend to procure the required apparatus. Three lights have already been ordered from France.[2] For the other eight we ask your kind offices with Her Britannic Majesty's Government, in order to get the apparatus through them. We shall give orders about the purchase money as soon as estimates are made out.'

Sir Harry Parkes having conveyed to Lord Stanley, then Foreign Secretary in London, the desire of the Japanese Government, he (Lord Stanley) laid the matter before the Board of Trade, telling that Department the Foreign Office concurred in Sir Harry Parkes' wish that the apparatus for eight lights be procured and sent off with as little delay as possible.

The matter was the subject of considerable discussion between the Board of Trade and Trinity House,[3] both departments agreeing that it was premature to obtain apparatus before a proper lighthouse service had been organised. It was finally agreed in June 1867, to refer the whole business to Messrs D. & T. Stevenson, Engineers to the Scottish Lighthouse Board, Edinburgh.

The Foreign Office in London received in August 1867 a further dispatch from Sir Harry Parkes requesting that the apparatus for five more lights, making thirteen in all, be dispatched, but as no money had yet been received, instructions were given that no expense should be incurred until Sir Harry Parkes reported on this matter.

The difficulty was removed in December by the receipt from the British Minister of a bill on the Bank of England for £10,729, and a promise of another similar sum in three months time. In the accompanying dispatch, he gave some further information regarding the positions selected for the lights and stated that a mixed commission of naval officers, in which the Japanese would join, had been appointed to fully consider the question. The report of this body organised in Japan having been received on the 14th of February, 1868, Messrs Stevenson proceeded to construct the apparatus required.

The fact of Japan's being subject to violent earthquake shocks raised a great sense of difficulty. The means thought necessary by them to avoid damage from seismic disturbances had the unfortunate effect of considerably lessening the efficiency of the apparatus designed for Japan. Metal reflectors were adopted instead of the much more powerful glass lenses, and the lanterns were made lower than usual, thus preventing their proper ventilation. A device was adopted of placing the apparatus on a table, which rested on balls, making a break in the rigidity of the structure, or what was termed an aseismatic joint, but which was an entire failure in practice. These precautions were in truth entirely unnecessary, as precisely similar lanterns and apparatus are erected in Europe, and have since been established in Japan by me, without their being damaged by earthquake shocks.

In reference to the effect of subterranean disturbances on such buildings as lighthouses, it is somewhat curious to know that the slight elasticity of the materials of which they are constructed entirely absorbs the motion of the foundations by the time it reaches the summit. Messrs Stevenson in their report to the Board of Trade on this matter said, 'It is evident that any lateral motion of the earth on which a building rests must be communicated to the foundations of the structure, and thence through all the rigid and unyielding material of which it is composed to its very summit, where the violence of the shock will be aggravated by the greater elevation of the building above the source of motion'. Precisely the reverse of this occurs actually, and Messrs Stevenson, accomplished and learned as they were, came to a wholly erroneous conclusion.

I had been on the balcony of a brick lighthouse, eighty feet from the ground, and had seen people immediately below me rush from their houses in panic on account of an earthquake shock, which I did not feel in the least. While the foundations vibrated, the summit remained perfectly quiescent, owing to the elasticity of the brickwork of the tower.

The Board of Trade wrote further to Messrs Stevenson, placing in their hands the 'selection of suitable persons to undertake the design and construction of the Lighthouses, and the introduction of a Lighthouse service in Japan'. It continued: 'If you cannot get engineers who have been trained to Lighthouse service, as the number of these is small, the Board of Trade thinks an active and intelligent Engineer with a general knowledge of his profession would soon, under your training, acquire such knowledge in these matters as would be sufficient. When he is selected, he and his assistants should proceed as soon as possible to Japan, and visit the sites proposed for the lights, design the buildings, and send home sufficient plans and data to enable you to design and construct the apparatus. It would also be very essential that one or two experienced light-keepers should accompany the engineers. A copy of the terms offered by the Japanese Government is enclosed for your information.'

The Messrs Stevenson, considering that I fulfilled the requirements, recommended me to the Board of Trade and I received the appointment on the 24th of February, 1868.

I at once entered upon a course of study of lighthouse work in Messrs Stevenson's office in Edinburgh and also visited many lighthouses and lightships at different points along the coasts of the United Kingdom, obtaining by these means a practical knowledge of their details and working.

Accompanied by two assistant engineers, Messrs Blundell and MacVean, I left Southampton on the 13th of June, 1868, and arrived in Yokohama on the 8th of August, 1868.

CHAPTER TWO

THE FIRST TELEGRAPH
IN JAPAN

BY THE LATE SUMMER of 1868 the Mikado's Government had firmly established itself in the city of Yedo, which was re-named Tokio. The opposition which still showed some vitality in the North had not yet been quelled.

As the work of organisation of the Government Departments, and the selection of suitable men to preside over these had, however, scarcely been begun, I reported myself to Sir Harry Parkes. This man of the hour met me with the greatest cordiality and threw himself heart and soul into the work. For many months he was the only medium of communication between the Tokio Government and myself and staff, conducting also all the financial operations. He at once enlisted the interest of the British Admiral [Sir Harry Keppel],[1] on the station, who not only assisted with his advice, but offered the aid of Queen Victoria's ships, should any be required. It was through Sir Harry's energy and unflagging interest that many, many months of delay in inaugurating the work were avoided.

The preliminary work being thus accomplished, decisions as to which points on the coast were most urgently in need of illumination, and the order in which the lighthouses were to be erected, were quickly made. The Lighthouse Establishment, consisting of dwellings, offices, stores, workshops, etc., was to be located at Yokohama.

The Japanese by themselves were at that time unable to carry out any large building operations. Indeed their skill in these was apparently the weakest of their accomplishments. It was therefore decided to obtain from England representative artisans in the different building trades, such as masons, plumbers, machinists, etc., besides several lightkeepers.

About the middle of October 1868, Mr Terashima[1], who spoke English and had strong foreign sympathies, was appointed one of the Governors of Yokohama.... Acting also in the Foreign Office, he was an able intermediary between the Imperial Government and the representatives of the Treaty Powers. As the charge of my duties was relegated to this officer, Sir Harry advised me to obtain Terashima's sanction to all work before carrying it out.

Being one of the first foreigners resident in Japan who possessed any technical knowledge, my assistance was eagerly sought for from the most diverse directions. I had come out as a lighthouse engineer, but I became unexpectedly the builder of the first telegraph line in the Far East.

It was decided to build one line of wire twenty-two miles long between Yokohama and Tokio, and to stretch another between the newly opened port of Osaka and the city of Kyoto. I was instructed to order the necessary material and appliances from England, and to obtain the services of an expert to execute the work and also to give instructions to Japanese as to its equipment and manipulation.

The material arrived in September 1869, and the lines were successfully erected under the superintendence of Mr G. M. Gilbert, who initiated the Japanese operators in their routine of work. Beyond a few of the poles being slashed by fanatical Samurai who must find some use for their swords, there was no evidence of any hostility on the part of the people. Messages were sent in both English and Japanese, the latter requiring forty-two characters.[2] A dual instrument was therefore adopted. Though presenting some initial difficulties, the use of dial and key was soon mastered by the native operators. The lines were in working order by the early part of 1870. [The first message was transmitted on the 7th January, 1870, and on the 26th the line was opened for Japanese telegraphy.]

I found, however, that though I was really the first to establish lines of telegraph in the Mikado's Empire, others in advance of me obtained sanction to do so. On hearing of the English lines being erected, the Swiss Consul informed Sir Harry Parkes that the Shogun's Government had given through him a concession to erect telegraphs all over the country. Presumably the Swiss Government, during the years of the Revolution, 1868-1870, had thought it well not to take any active steps until it saw 'how the land lay'. Sir Harry Parkes, having a more accurate knowledge of the trend of affairs, was enabled to forestall them. It was averred that the proposal of the Swiss Government was to work as well as to erect the lines. Diplomatic difficulties arising on this point, the Japanese, after 1868, never permitting any foreign control whatever, the agreement was eventually dropped. The Austrian Government had also previously supplied a complete set of telegraphic apparatus with a large quantity of wire. This was discovered in one of the Shogun's warehouses and was afterwards erected by the author's staff between the Imperial Palace and various Government offices in Tokio.

The two pioneer lines of telegraph, about one hundred miles in length, formed the nucleus of the extensive system of today.[32] They were placed under a separate Government Department, and were later merged into the Department charged with railway construction.

CHAPTER THREE

LAYING OUT A NEW
SETTLEMENT

BESIDES the treaty of 1866, another agreement[1] was negotiated and signed in the same year, with the object of making the settlement of Yokohama a place fit for the residence of Europeans and Americans.

Few persons who have not visited Japan can form any conception of the primitive character[2] of the average dwelling. The typical house is both a model of simplicity and a pattern of discomfort in all seasons. Upright posts which rest on stones, the tops of which are slightly elevated above the level of the ground, form the main structural feature of the edifice. These carry an enormously heavy and badly constructed roof, which is covered with heavy tiles or a great thickness of thatch.

The first floor is formed about eighteen inches from the ground, leaving an empty space below, which is generally a receptacle for all kinds of rubbish.[1] Between the upright posts is placed a slight lattice framework of wood, on which the native paper of a beautiful silky texture is stretched, forming the partitions which divide the houses into rooms. These *shoji*, or screens, slide in grooves at top and bottom and are easily removable. There are no means of artificial heat except for braziers, or *kotatsu* (fire-places sunk in the middle of the room for red-hot charcoal), or any means of ventilation except by removing a screen, or through the upper lattice work dividing the rooms. Glass is not employed, the only light available being diffused through the paper on the screens. While in a few houses there are second floors, access to which is gained by means of a ladder, they are generally only one-storied.

These houses built in lines, form between them what may by courtesy be called a street. On each side of this, within three or four feet of the house fronts, is a stone-lined drain, which is sometimes covered by loose planks of wood. As, in many cases, this drain has no slant, it becomes in dry weather a reservoir for stagnant and dirty water. At the back of the houses there are generally two carefully protected casks sunk in the earth forming cesspools. Human excreta, used for manuring the paddy, or wet rice fields, are regarded as being of great value.

Hard and dry roads were not thought of in old Japan. There was no horse traction, and only a little wheel traffic by push carts. Both man and horse wore sandals made of straw. Human beings used wooden clogs which raised their feet an inch or two above the ground and protected from the mud. The streets were formed by simply levelling the earth down; nor was any further interference with the virgin soil attempted. The flattening of the surface was generally done in a most perfunctory manner. After a heavy rain, the streets presented the appearance of a body of water rich in islands, much like their own Inland Sea.

Arriving in Japan, I found the foreign quarter of Yokohama precisely in the condition described above, though of course the houses of Americans and Europeans, usually 'bungalows' or one-storey tile-roofed houses, were constructed somewhat after the fashion in Western countries.

The agreement of 1866 provided that all roads in the settlement were to be laid with broken stones, or gravel, and 'metalled'. It also stipulated that sidewalks should be formed, the open ditches be removed, an organised system of surface drainage be carried out; and that those portions of the ground too low for drainage purposes were to be artificially raised. An adjoining eminence, now called 'the Bluff',[3] was to

31

be let to foreigners, upon which to build residences. Additional land might be acquired should the settlement extend beyond the area originally contemplated.

Sir Harry Parkes lost no time in urging the local authorities to utilise my services in transforming the settlement from its unhealthy and disagreeable condition to one more suited to the requirements of Western civilisation.

The necessary plans and investigations having been made, I presented my report. A copy was also presented to each of the foreign representatives. On their approving it, and the Governor having satisfied himself of the probable cost of executing the works proposed, I was instructed to proceed.

This was a task of considerable magnitude, embracing as it did the formation of many miles of macadamised streets, side paths, surface drains and an underground system of pipes. As nothing at all of a similar nature had ever before been attempted in that part of the world, this was also an enterprise of much interest.

The difficulties at first seemed almost insuperable. In the first place, no stone was to be had within one hundred miles of Yokohama. When found, if the quarries were not on the sea coast, the only means of transporting the stone was by packhorses walking on footpaths, there being neither broad roads nor carts. In this country of bamboo, such a thing as an underground drain pipe of baked clay had never been thought of, though good pliable potter's clay was known to exist near Tokio. Such iron work as was required could, however, be obtained from foreign engineers who had settled in the place, principally with the object of ship repair.

After searching long and far, a moderately hard stone was obtained near the harbour of Shimoda,[4] over thirty leagues distant. A fleet of native boats was at once chartered to bring what was required of this to Yokohama.

Having explained to the officials the nature of a drainpipe, and provided them with full-size drawings of them, I was one day agreeably surprised to have a sample brought to me which I regarded, though insufficiently burned and soft, as a promising production. This was afterwards much improved on and arrangements were made for a supply to the quantity desired. After some hundreds of this new sort of ceramics had been laid and the ground filled in above them, it was discovered that they were not equal to the burden laid upon them. From the pressure of the superincumbent earth, every drain tile was found cracked and splintered in every direction and had to be removed. Eventually, a satisfactory pipe was produced, which served its purpose efficiently.

The next problem was to procure a roller to level the broken rock and give it a 'metalled' surface. A stone weighing four or five tons was obtained from the quarries and rounded into shape, making a heavy roller able to consolidate the paving material. Next horse-gear was bought, but where were the animals that were used to harness? Not a single beast in Japan, except oxen, was accustomed to traction, the horses[5] being used only to packs and saddles. The only course, therefore, was to fall back on native methods and employ human beasts of burden. For nearly a year, a gang of twenty or thirty *ninsoku*, or labourers, dragged the stone roller up and down the new streets of Yokohama.

The delays in obtaining material, both stone and drain pipes, as well as the novelty to the Japanese of the whole proceedings, rendered execution of contract tedious in the extreme. Nevertheless in the autumn of 1870, the work was completed to the satisfaction of both foreigners and natives.[6]

CHAPTER FOUR

WATER AND LIGHT

HAVING GOT clean hard roads and a suitable system of surface drainage, desires were expressed on all hands for a water supply and for some method of lighting the streets. Neither of these having been mentioned in the agreement of 1866, the Japanese Government refused to bear the expense of carrying out these costly propositions.

The dangerous hygienic conditions of the water supply[1] at the time were fully recognised by both aliens and natives. As the houses drained into cesspools not far away, the earth soon became impregnated with poisonous matter. The wells from which the water was obtained were also sunk at the back of the house, in more or less proximity to the cesspools. The risks of contamination of drinking water were therefore very imminent and real.

As early as April 1869, Terashima informed me that the native merchants had subscribed $250,000 in order to supply Yokohama with pure water, and that they desired me to form a scheme for the work. One year later, after having searched in many quarters for water, and finding what I considered to be a suitable source, I presented my report to the Government, but my proposals seemed altogether too extravagant for the Japanese mind of that day. I suggested a filtration plant and also a reservoir, but for neither of these could they see any necessity.

In my idea water was to be conveyed in iron pipes by gravitation, but they desired to carry it according to a method very old in Japan, that is by means of trees hollowed out in the centre and through bamboo tubes. The water was to be raised at necessary points along the route by a series of buckets working by hand on revolving ladders.

Nothing therefore was done towards carrying out this scheme at the time, though many years later a much more elaborate plan was carried out under the supervision of a British engineer.

As regards the lighting of the streets, although the Government, with the object of meeting the needs of the foreign community, put up lamp posts, nothing was subscribed towards the cost of maintaining the lights,[2] and the matter was postponed until gas was introduced, many years afterwards.[1]

CHAPTER FIVE

BUILDING IRON BRIDGES

ANOTHER WORK, illustrating the spirit of progress which then actuated the Japanese was undertaken by me at this time, viz., the building of an iron bridge.[1]

The type of bridge to be seen in Japan in 1870 was, like that of the dwellings already described, of a very primitive character.

The piers were formed by two trees with the bark on. These were driven into the ground as far as native appliances would allow. The space between these was then spanned by two other tree trunks, selected as having the necessary bend which gives Japanese bridges their arched form. On top of these wooden planks were laid crossways. A rough handrail completed the structure. Such bridges were in need of constant repair, were utterly unfitted for the traffic of vehicles, and had to be rebuilt entirely in about every five years.

When Terashima was expected to re-erect one of these short-lived structures, which should carry the main road from Yokohama to Tokio, he broke away from routine, consulting me as to establishing a bridge of some permanent character. He explained that its construction would be in the nature of an experiment, and that he desired to show his countrymen how bridges were built in Western countries. He had, however, no authority to spend any large sum of money, and could not think of ordering one of iron from Europe. Moreover, he objected to the employment of foreign artisans already experienced in such work.

On learning these decisions, it became a question, whether it would not be more prudent for me to give up the whole matter, or to undertake the work with such assistance as I could obtain on the spot.

Having had considerable experience in the construction of railway bridges in England, I somewhat recklessly decided on the latter course.

Making a design for a simple iron lattice bridge with stone abutments, having about one hundred feet span, and carrying a roadway twenty-five feet wide, I called to my aid a fellow-countryman.

Happily there was an Englishman in Yokohama who knew a little about blacksmith work, and I at once obtained his assistance. The foreign settlements in Japan and China were then scoured for iron plates, but the bulk of those I secured came from Hong Kong.

A punching and shearing machine was borrowed from an engineering shop, by means of which the iron plates were cut to the required sizes and holes were made for the rivets. The whole girders were thus fitted and riveted together by Japanese mechanics who had never in their lives handled similar tools, and were completely ignorant of the exact character of the work in which they were having a part.

This piece of work evoked the greatest interest in the public mind. During its construction, crowds of people from both the samurai and the mercantile classes came to see the To-jin's[1] new-fangled project. The men sat on the ground or on the large stones used in the building operations, smoked their little pipes, refilled and smoked again, and rolled their tiny pellets of shredded tobacco, lighted and puffed again, while discussing among themselves, and with many a *naruhodo*, what to them was a curious novelty. I afterwards gleaned these wiseacres seriously doubted the efficiency of the whole enterprise.

But 'nothing succeeds like success', especially in Japan. The first iron bridge in Japan was erected. When tested to considerably over its full working load, except that the looseness of the riveting allowed it to deflect at the centre slightly more than it should have done, it appeared to be perfectly satisfactory.

CHAPTER SIX

OSAKA: AN INTERNED CITY

ON MY FIRST visit along the coast, in December 1868, I was requested by Sir Harry Parkes to see the governor of Osaka, one of the three *fu*,[1] or Imperial cities, and the principal emporium of trade in the central portion of Japan. This was one of the new ports which had been opened to foreign trade at the beginning of the year.[2] It had excellent internal communication with the interior by means of a large and deep river with various branches, flowing through large and productive districts of the country. This river, the Yodo, receives the overflow of Lake Biwa and drains a large watershed. It intersects the town and enters the sea a few miles distant. At its mouth, however, was a sand bar which did not permit of the exit of any but flat-bottomed barges, or small Japanese punts. On its dangerous rollers, hundreds of boats had been upset and thousands of lives lost.[1] The efficiency of Osaka as a port of foreign commerce was therefore wholly destroyed. As the river was the only means of getting on board ships from old Japan's chief commercial centre, the existence of this sand bar apparently nullified the long and wearisome negotiations and infinite labour from which its opening resulted. Nature seemed to have been victorious over diplomacy.

Anxious, therefore, to have some professional opinion upon a possible amendment of this state of affairs, Sir Harry Parkes urged Governor Godai[3] to allow me to make a report indicating what could be done. Happily Mr Godai, born in Satsuma, who had been in the Foreign Office in Tokio, but left Government service for private business, was a man with advanced ideas. He at once displayed great anxiety to have the river clear for the passage of ships. He sailed with me over the bar, but explained that the local Japanese authorities had a scheme of their own, which they felt sure would be effective, and which he was trying to get the Imperial Government to carry out.

Deeming this plan quite impracticable, I proceeded to take soundings and obtain such information as was necessary to formulate a definite proposal. The land alongside the river was perfectly flat and raised only one foot or two above the water. My idea was to cut a new channel through the soil, diverting the inland waters and the accompanying silt through this, dredging a channel through the existing bar and utilising the large tract of deep water inside as a harbour. I expected to use the earth removed by digging to make banks for the new river.

Having matured this scheme before I left Osaka, I handed a copy of my plan and report to Mr Godai with the expectation that he would consult the Imperial Government concerning its practicality. Sir Harry deemed the scheme feasible, but had some doubts of the propriety of his pressing the Government in Tokio to incur the large expenditure involved. He, however, brought it before the notice of the Foreign Office. The Mikado's servants promised to give it their consideration, but protested that the Governor of Osaka had not consulted them in the matter, nor did they know that he had been furnished with particulars of the scheme. At their request, additional copies of my report were furnished to them.

Nothing further was heard of the matter for some months. Then the British Minister obtained information that the work was being executed, and, so far as could be judged, on the lines I had proposed. He was told that a River Department of the Government had been formed, and it had accepted my scheme, and was carrying it out on their own initiative.

As this matter was of such moment to the whole foreign community in Japan, and as it was so difficult to obtain any trustworthy information as to what was being done, it was arranged that Mr Blundell, one of my assistants, should visit the locality and report upon the circumstances. Osaka was rapidly increasing in proportions and population.

Mr Blundell found that considerable progress had been made with the new channel proposed by me, but that the excavated earth, instead of being made a bank on each side as intended, was carried a long distance away by a tramway, the rails and rolling stock for which had been purchased from an enterprising English merchant for the respectable sum of $50,000. At the scene of operation great excitement was apparent, when the object of Mr Blundell's visit was made known. Warned by a mob of labourers not to approach, he was, on retiring, treated to a fusilade of stones, narrowly escaping serious injury.

It was learned also that the 'River Department' and Governor Godai were at variance as to the way in which the work was being carried out, which, as the latter averred was incorrect and would result in failure. Mr Godai was desirous of obtaining my opinion as to the practical value of the method adopted, but the 'River Department' would not listen to the suggestion. Therefore as Governor of the Imperial Fu, or City, he exercised his authority and stopped the work. After having fruitlessly spent probably about $250,000 upon the attempt to carry out the scheme without consulting me, the whole enterprise was abandoned.

As a seaport, therefore, Osaka[2] has remained completely ineffective, but the railway now constructed between that town and Kobe,[4] which has a good anchorage, conveys all the traffic of Osaka to sea-going vessels.

CHAPTER SEVEN

TAMING THE RIVERS

THOUGH occurring some years after the events just recorded, it may be convenient at this point to state what took place in regard to another of Japan's great rivers, as showing the peculiar methods of the Government at this period of its existence.

The treaty port of Niigata,[1] nominally opened to foreign trade on the 1st January, 1868, is situated at the mouth of the Shinano-gawa, or river. As at Osaka, the passage into this river is blocked by a great bar of sand which had only five feet of water over it; the Sea of Japan being in this latitude quite tideless. The anchorage outside was unsheltered, making the loading or unloading of vessels moored in it quite impracticable. Nevertheless inside the bar, for a mile or two up the river, there was a depth of water of from eighteen to twenty feet. The existence of Niigata as a trading port depended, therefore, upon the removal of the obstacle at the mouth of the Shinano River.

As a matter of vital interest to the foreign trading communities in Japan, Sir Harry Parkes urged the Imperial Government to take some measures to make Niigata valuable as a port. At the British Minister's instigation, I was instructed to visit the place, examine the river, and make a detailed report.

I arrived in June 1871 on the West coast opposite Korea, and spent some weeks in a survey of the conditions. Let me here quote in substance from my paper read in November 1874 before the Asiatic Society of Japan at one of its meetings in Yokohama:

'The Shinano-gawa is the outlet of the drainage waters of large tracts in the Provinces of Shinano, Echigo and Musashi. Its general course is northerly and its total length is about 250 miles. Its discharge, as gauged by me while there, was 1,500,000 cubic feet per minute, its summer flow being calculated as 700,000 cubic feet per minute, and its flood discharge 14,000,000 cubic feet per minute.' These figures were in 1873, quite independently, verified by Lieutenant J. A. Lindo of the Royal Engineers in the Dutch Army, whom the Government had then taken into its employment. As the Thames discharged 400,000 cubic feet per minute and the Shannon in Ireland, the largest river in the British Isles, only 5,000,000 cubic feet per minute in flood, some idea may be obtained as to the relative size of this Japanese stream.

For forty miles up the river erosion of its banks has widened out until in some places its bed is three miles in breadth, and, over the whole of that distance, shallows and sand banks have been formed which not only impede the flow of water, but almost entirely destroy the use of the channel for navigation purposes; the depth of water in the channel being in places not more than three feet.

The banks, formed of a fine sand, and very low, were damaged by every flood, and I was informed that 12,000 acres of valuable land were covered by water five or six times each year.

Instead of raising the banks, or otherwise regulating the flow of the river, the scheme to remedy this state of matters, which had been adopted by the Japanese authorities, was the formation of a new channel of huge dimensions to carry off the surplus water. The river running for some distance parallel to the seacoast and about eight miles from it, this huge cutting, in some places over sixty feet in depth, left it at a point named Okodzu, forty miles from the mouth, and proceeded through the eight miles of intervening space, direct to the sea, at a place called Taradomari, about twenty-six miles south of Niigata.[2]

This scheme, which is an extraordinary instance of perverted intelligence, would have had the undoubted effect of still further silting up the mouth of the channel at Niigata, by preventing scour in times of flood.

The bar, which presented no peculiar difficulties to hinder its being successfully dealt with, had, on the contrary, many advantages which, with proper treatment, rendered its improvement a comparatively easy task. It was composed of the finest sand, which was moved about by every disturbance in the water, whether by waves from the seaward, or by river currents from inshore. It was stated that this bar consisted of at least thirty feet depth of this fine sand. With a properly directed stream, carried through a channel of a width necessary to get a maximum scour in floods, this bar could not fail to be swept away.

I felt convinced from the success of the works at the Sulina mouth of the Danube, the Oder, the Tees, and other rivers, which were, in many ways similarly circumstanced, that the means adopted at these places would have the same result here.

My scheme, in contradistinction to the extra-flood channel of the Japanese, therefore, was simply to confine the river flow by means of wooden piers carried outward to the sea, to such a width as, while allowing for the free discharge of floods, would secure a maximum scour until deep water was reached. The effects of such piers at the Sulina mouth of the Danube, given by Sir Charles Hartley, in the Proceedings of the Institution of Civil Engineers, was then stated. I knew no reason why similar results should not follow the erection of such piers at Niigata; that is, the creation of a deep channel kept clear by its own scour.

While engaged in making my investigations at Niigata, I explained to the Governor of the District the principles which should guide me in any scheme I might lay before the Government in Tokio. These, involving the cessation of work on the new cutting, were stoutly opposed. Even their accuracy was impugned by the native officers. It was averred that nearly a million dollars had already been spent on the scheme inaugurated, and that the Government would certainly complete it. To this I responded that, if such were the case, further investigation by me was useless, and that I should return to Yokohama. But after patient discussion, withal being enabled from books in my possession to give evidence of scour having the effect of deepening river channels, light began to dawn on their minds, and I was eventually requested to proceed with the formulation of my scheme.

On completion, I laid my plans, reports and estimates of cost before the Government in Tokio, and, also handed copies to the British Chargé d'Affaires (Sir Harry Parkes being absent on leave at this time) on the 8th of July, 1871.

These papers dealt with the mouth of the river and the formation of a harbour only, but the state of its course for forty miles up (already described as a waste of sand three miles wide in places) was pointedly referred to in the report. The non-completion of the new cutting was strongly insisted on, and instead it was recommended that, by artificial banks covered with wicker work or other material to prevent erosion, the water be confined to a natural width.

In order to form a natural judgement on the cost of such work, it was pointed out that a survey of the river for fifty miles up with sections and other information should be obtained.

On the 22nd of August, or six weeks after my first report had been delivered, I received instructions to undertake this survey and inquiry. My two assistants, Messrs Fisher and Wilson, were directed to perform the work. They occupied three months of unceasing labour, and made a complete plan of the river and adjoining country for a distance of fifty miles from the mouth of the Shinano River, with cross sections, showing its depth, at frequent intervals.

On their return, a second report, with drawings fully presenting matured proposals for the work of confining the river within its natural width, for reclaiming the lands destroyed by the floods, and for preventing further overflow, was delivered to the Government on the 20th of March, 1872.

No response was given to these, and Niigata seemed doomed as a trading centre. The enormous river, which if properly treated, would have made the city accessible to sea-going ships, was allowed to spread itself over the same vast space as before. The Chamber of Commerce of Yokohama passed resolutions to the effect that, as no communication could be made with Niigata, and as the Government were doing nothing to remedy matters, the port had better be closed entirely.

With his usual vigour, Sir Harry Parkes made another attempt to rescue the place for trade. Shortly after resuming his duties in Japan at the beginning of 1873, he wrote to me as follows:

18th April 1873.

Dear Mr Brunton,

At an interview I had yesterday with two of the Councillors of the Government (Okuma being one) I brought up the question of the Shinano-gawa and ascertained that they contemplate sending a Dutch Engineer to survey the whole course of the river.[3] I asked whether this measure was being taken with your knowledge, and as a continuation of Mr Fisher's work. They did not seem to know that Mr Fisher had ever worked up the river, but thought he had limited his survey to the mouth only, and his work to the harbour question solely. I undeceived them on this point, said that you had showed most clearly that the river and harbour question were one, and that Mr Fisher had made a survey of the river from the mouth to Okodzu, some 45 or 50 miles, and added that they should at least consult you before they changed their men and plan of operations. I also remarked upon the importance of the petroleum springs and the interest you had taken in that matter, and urged immediate reference to Yamao[4] [Vice Minister of the Public Works Department, which had then been formed] in order that all you had urged in regard to the river, and wished to urge as to the petroleum, might be duly considered.

I think you should know of this conversation in order that you may judge whether it would not be worth while to obtain an interview with Yamao and

45

talk the subject over. I do not know in what light you would regard the work being placed in other hands. If you have enough to do already you might not care about the transfer, but, otherwise, you might not wish to see another man entering into your labours, and Mr Fisher's work thrown away or absorbed by another person.

I should mention that I placed in Okuma's hands your second report in order that he might satisfy himself as to what you had done, and at what point the question stood when you left for England.

The discussion will, at least, enable you to enter on the petroleum question and decide whether you should go to Niigata to meet the *Thabor* [the Lighthouse tender], if you wish to bring these questions up.

I should like to know what you propose to do.

Very truly yours,
H. S. PARKES

I visited Yamao, but obtained no satisfaction. I regarded the Niigata episode therefore as closed. The labour and money expended on making investigations of a most complicated character seemed to be thrown absolutely away. As late as 1875, Okuma, then Minister of Finance, informed me that it had been decided that my scheme was to be carried out, under my own supervision. But it appeared that Okuma's financial proposals did not, at that time, meet with approval; and finding himself in a minority, he resigned, and the great works he had in contemplation were indefinitely postponed.[1]

CHAPTER EIGHT

THE GOLD MINES OF SADO[1]

WHILE AT NIIGATA I paid a visit to the island of Sado, situated about thirty miles due west from Niigata, and greatly celebrated in the annals of Japan for its gold production.

About thirty miles in extreme length, and fifteen miles in width, Sado Island is said to be extensively permeated by auriferous quartz. Its most lofty eminence rises to a height of 4,500 feet above the level of the sea, and its surface generally is irregular and precipitous in character.

I found two Europeans on the island. One was Mr Erasmus Gower, a mining expert who had introduced the use of explosives in mining[21] and the other Mr McNichol, a mechanical engineer and an all-round useful man. This gentleman had laid a line of tramway from the mines then worked to the shore. The rails were formed of baulks of wood having strips of iron fastened on them, and the rolling stock, including the wheels and axles, was made on the island. He was at the same time engaged also in building a small steamer, making all the necessary machinery on the island in order to obtain a more convenient mode of communication with the mainland than the native boats afforded.

A very considerable amount of milling machinery, erected in sheds, driven by a large stationary engine, and consisting of stone-breaking, crushing and amalgamating appliances, necessary for the recovery of the gold, was at work, and in excellent order.

Accompanied by Mr Gower, I inspected the mines

then being worked. They were twenty-nine in number, and four thousand persons were employed in them. In the neighbourhood there were hundreds of horizontal adits or tunnels, about seven feet square, piercing the faces of the hills to various distances. These old mines had been operated during various periods in the preceding centuries, but were now abandoned.

In other districts the island is honeycombed with ancient workings, of which there is no record as to when they were made. Huge heaps of quartz, judged by the natives to be insufficiently rich in gold to pay for treatment, were piled up in every direction. Yet Mr Gower felt convinced that with improved appliances, reduction of these ores would be quite remunerative. The amount of gold in the quartz is very variable, but the Japanese had only attempted to deal with richest portions.

Unfortunately, I had at this time no correct figures showing either the proportion of gold per ton of quartz, or the amount obtained per annum. I have, however, Mr Gower's general assurances that the great body of the quartz is exceedingly rich, that is, so far as can be judged, and in enormous quantity, and that a very considerable value of gold was then being extracted annually.[2]

CHAPTER NINE

THE PIONEER RAILWAY IN
THE FAR EAST

AS EARLY AS February 1869, Japanese officers expressed themselves extremely anxious to introduce railways into Japan.

I have reason to believe that my descriptions of the speed, ease and cheapness of railway travelling greatly stimulated this desire. Yet while raising no serious objection to the making of railways, I had formed an opinion, which I repeatedly expressed with great emphasis, that the immediate and pressing need of the country was not so much an elaborate and costly railway system, as the formation of good roads. Besides the main thoroughfares between Tokio and Kioto, themselves merely muddy travelling ways, almost impassable in wet weather, the only tracks by which journeying could be accomplished or merchandise conveyed were narrow footpaths forming the dividing ridges between the irrigated fields. It was only by pack horses, walking in single file, that the products of one part of the empire could be transported to other parts. The development of the country, in a manner at all commensurate with the desires of His Japanese Majesty's Government, would be seriously impeded until a remedy was applied to this discreditable state of affairs.

It appeared to me that the energies of the country would be more suitably expended on making good public roads. These would become feeders for the railways which must in due time follow. I pointed out that a government can bestow on a nation no greater benefit than the improvement of the common highways. I also quoted the opinion of a highly respected authority[1] who remarked: 'Let us travel over all the countries of the earth, and whenever we shall find no proper facilities in travelling from a city to a town, or from a village to a hamlet, we may pronounce the people barbarians.... A point to be particularly borne in mind in circles betraying a tendency to railway mania is, that common roads are not so much the proper supplements of railways as railways are of common roads, and it can only be where local conditions are very exceptional[1] that railways are entitled to precedence. To begin with railways before there are roads is generally to begin at the wrong end.'

But my representations were of no avail. The modest and comparatively simple work of forming a series of roads raised no interest in Japanese minds. They desired that more heroic method of developing the country which railways seemed to offer them.

Previous to accepting an appointment to Japan, my whole professional career had been in connection with railways. I was therefore quite familiar with the details of their construction, and was able to give the Japanese any information regarding them which they might require.

When asked to make a report upon a line between Yokohama and Tokio, I at once set to work at plans and specifications. In the middle of March 1869, I presented these to the Imperial Government. I pointed out that no constructional difficulties were likely to occur in executing the work, and that its probable cost would be $5,000,000.

When, however, financiers heard of the possibilities of a lucrative investment of their capital, they went up to Tokio to obtain concessions. Mr McClelland from Bombay was the first to ask for these, but his proposals met with no success. Mr H. N. Lay,[2] formerly Comptroller of Customs to the Chinese Government, who had been superseded by Mr now

Sir Robert Hart, arrived in Japan about the end of 1869. He at once made some very high-sounding proposals, offering the Government $5,000,000 for the concession to construct and equip a railway between Tokio and Kioto, announcing that money for the purpose had already been subscribed. Apparently his plans were concurred in by the Mikado's Government, or at least he thought they were. Leaving Japan with his proposals, he returned to London. There he discovered that the cash had not been subscribed. When he attempted to raise the money, he experienced the greatest difficulty. Chagrined at the course which events had taken, the Imperial Government of Japan withdrew from their understanding.

It was then decided to make two lengths of railway: one of twenty-two miles from Yokohama to Tokio, the other of twenty miles from Osaka to Kobe. After completion and trial of these, the construction of others, including trunk lines, might follow. The financial and purchasing arrangements were placed in the hands of the late Oriental Bank, London, whose directors appointed and employed the European staff required for the lines.

Mr W. W. Cargill, formerly a veteran manager in the Bank's service, and with experience in India, received the appointment of General Manager, and Mr Morel, 'a gentleman of experience and ability', was appointed to be Chief Engineer. A large staff of clerks and artisans was also appointed.

Work on the line to Tokio was begun in the middle of April 1870. The rails were laid to a gauge of three feet six inches, and the line was single. Its bridges were temporary wooden structures. The rolling stock consisted of small tank engines, obtained from England, and corridor carriages. The general outward appearance of the passenger and 'goods', or 'freight' trains, was that of English rail transportation.

Eventually opened to the public in the autumn of

1872, trains were run every hour from each terminus, occupying one hour on the journey.

The making of this line, the pioneer of railways in the Far East, was, perhaps, not unnaturally, attended by a series of the most unfortunate mischances and mistakes.[3] Buildings were erected, pulled down, and re-erected in other places; numerous diversions were made; bridges were strengthened after completion; rails were twisted in every conceivable form and laid in such a way that it seemed impossible for a train to run over them. The total cost of the road was reported to be fabulous, but no official statement was given of the amount of this.

The main cause for this somewhat deplorable condition of affairs was that the European staff engaged to direct operations, headed by Mr Cargill, the General Director, supinely permitted the interference of the native officials with their operations.

Wholly unacquainted with the system necessary for the proper and economical demonstration of such a work, but self-willed, self-satisfied and overbearing (when allowed to be) the Japanese required to be firmly led, as was clearly exemplified by the Railway Works, in order to avoid a complete disorganisation of operations.

The Japanese at the present day have nearly six thousand miles of railway, which are worked satisfactorily.[2]

CHAPTER TEN

MAPS,[1] SURVEYS AND ENGINEERING EDUCATION

AS EARLY AS April 1869, Terashima, Minister of Foreign Affairs, applied to me in reference to the possibility of a complete survey of the empire. Strangely enough, what were judged to be fairly accurate maps of the country were already in existence when the foreign treaties were signed. These were on too small a scale, and were lacking in detail, but they depicted the course of rivers, the outline of mountains, and the situation of towns with considerable accuracy. In fact, so correct was the line of the coast delineated, that it was adopted by the British Admiralty for the charts, by which ships were navigated.[1]

This went to show that, at some time or other, people having a knowledge of trigonometrical surveying must have resided in Japan. These may have been the Portuguese or the Dutch, or the learning may have been obtained from Sidotte,[2] the Italian priest whose services were so appreciated in the early part of the 18th century by the Taikun of that period.[3] From whatever source the information was obtained, however, there was no trace of the possession of it by any native, so far as was known to me, when Europeans finally settled in the country.

These maps having been frequently brought to my notice, I thought they might be utilised to make a map of the whole of Japan on scientific principles and according to methods used in Western countries. One of my last acts, before finally leaving Japan, was the compilation from Japanese data of a map to a scale of twenty miles to the inch, which showed towns and villages, rivers, mountains, roads, paths, etc., in as great detail as possible. By the aid of interpreters, the names written in Japanese, were translated into Roman characters,[4] according to the system of spelling, which was then generally adopted, and which had been recommended by Mr E. Satow, Secretary of the British Legation and then regarded as the best foreign Japanese scholar in the world. This map, purchased by various Government Departments in Europe and by the commercial houses having dealings with Japan, was, after its publication, generally regarded as the standard work of the kind. Travellers in the interior also found it of value to them. Miss Bird[5] in her *Unbeaten Tracks in Japan* speaks of it as guiding her on some occasions and failing her on others.

The desire of the Government, as expressed by Terashima, was to obtain a fully detailed map of the Empire, something similar to the ordnance maps of Great Britain. I stated the advantages of such maps in the new Japan, now no longer feudal, but under Imperial rule, in which works of improvement, such as roads, railways, mines, water works and such like were in contemplation, but I felt obliged also to show clearly the magnitude of such an undertaking. In a memorial addressed to Terashima, I gave the various scales to which the Ordnance Maps in England were drawn, informing him that the production of these had cost £5,000,000 sterling, and that the annual expenditure on the survey at that time amounted to £100,000. I gave my opinion that to produce a map of Japan of the smallest ordnance scale, viz: one inch to the mile, would probably cost £2,000,000, or about $10,000,000. These figures, somewhat staggering in their magnitude, produced the effect which might be anticipated, and all idea of a survey of the country was indefinitely postponed.

Less ambitious surveying schemes were, however, under consideration, and some of these were rapidly carried out under my direction and supervision. The first was a survey of the settlement and town of Yokohama. This was completed in April 1870. The second, a survey of the Bay of Yokohama, with complete soundings, was conducted with the aid of Captain Brown[6] and completed about the end of the year 1870. This officer, late of the Peninsular and Oriental Company, was then in the service of the Lighthouse Department of the Government. The third was a survey of the vicinity of Yokohama to the extent of six miles around it, that is, the area within treaty limits. With the aid of two extra assistants, this was finished by August 1871.

Besides being of immediate use, these surveys had the effect of awakening in the Japanese mind a desire to attain the knowledge necessary to make such themselves. Orders were promptly given to me to procure from England theodolites, quadrants, sextants, eidographs and various drawing instruments. On the arrival of these, my services and those of my assistants were enlisted, to show the young men how to use them.

But not having been engaged as schoolmasters and having on hand other important work, which kept us fully employed, it was impossible for me and those acting with me to give this matter the attention it required.

Nevertheless, though not imparting much actual knowledge, we were successful in demonstrating clearly to the Japanese that the work of trigonometrical survey was not one in which they, without any preparatory education, were likely to become instantly efficient. I never ceased to impress on them the necessity for obtaining at least a rudimentary acquaintance with mathematics, before attempting or hoping for signal success in land surveying.

By November 1870, the Japanese, becoming alive to the accuracy of my representations, resolved to form a regular school for mathematics, and other cognate subjects. Under my guidance, a large building was erected close to my office in Yokohama, and such educated men as could be found in the ports in Japan or China were engaged as teachers. Appreciation of my services was expressed and those in authority voluntarily offered to increase my salary.

A great rage for education set in,[2] and pupils, young men of the Samurai class, were, under the auspices of the Government in Tokio, sent in batches of twenty and thirty at a time to my establishment. Though most of these lads devoted themselves to studying the English language, many were more ambitious. These aimed at the higher branches, expecting to master them speedily. While many of these practised mechanical drawing, tracing, and such like, others took advantage of the permission given them to learn navigation from Captain Brown on the lighthouse tender. Others confined themselves to the simple study of mathematics, being obliged to do so through interpreters.

Progress in real work was as erratic as a lightning flash. The desire was to learn everything in a hurry. Complaints were made by the pupils that the teachers were not pushing them on sufficiently fast with their studies. Appreciation of the difficulties in front of these men, who had chosen a royal road, became more felt every day. An outbreak of discontent ensued. As many as thirty and forty pupils were in the habit of absenting themselves, they being all alleged to be 'sick'. This was their way, and the traditional way in old Japan, of making a protest against some grievance.

Yet the work went on. Books and instruments in great numbers were ordered from England. Among other publications which I was requested to obtain were two copies of the complete Encyclopedia Brittanica. At the same time orders came from the Government in Tokio that a number of young men, selected from each bureau of the Department of Public Works, were to be sent to Great Britain to learn engineering.

One of these thus chosen was Mr Fujikura, my most valued interpreter, who, on his return to Japan, became my assistant. When Europeans were dispensed with, he was appointed as Engineer in Chief to the Lighthouse Department.

Meanwhile a series of large buildings had been in course of erection in Tokio, which were to form an Engineering College;[3] and competent instructors were selected in Great Britain to teach in it. Much to my relief and convenience, the schools of Yokohama, which had been organised under my supervision, were absorbed by this college under the direction of the Department of Public Works.

Having done so much in the way of education, and having further been much consulted about the character of the education to be afforded in the College, I tried to impress upon the men prominent in the Imperial Government the necessity of their adopting a system that should lean more towards practice than to theory.[7] Their desire was to have men who would be speedily useful in the work they contemplated, and my idea was to facilitate their laudable purpose.

Marine surveying aroused much interest among the Japanese, and a vessel was obtained with the object of accompanying H.M.S. *Sylvia*, then on surveying service in Japan. Captain St John,[4] commanding this vessel had, at Sir Harry Parkes' request, promised his assistance, and, as he was then engaged in making surveys of portions of the Japanese coast, the students would be enabled to watch his methods. I was instructed to obtain from England the instruments necessary for such work, which I did with Captain St John's kind aid and advice.[5]

Other establishments of a technical character, for promoting various other branches of education, were instituted by the Government at this time, but these, not coming within any special cognisance, I am unable to speak of authoritatively.

CHAPTER ELEVEN

THE NEW COINAGE

THE YEDO TREATY of 1866 provided for the establishment of a free Mint. It was decided to erect this at Osaka, and its establishment and direction were placed in the hands of an English staff, headed by Major Kinder,[1] who had been director of the Mint at Hong Kong. The new milled coinage of *yen* and *sen*, similar in value to dollar[2] and cent[1] coined in this establishment, soon superseded the old confused and debased currency used in feudal Japan. Under its able directorate, coins were struck which when proved by the Mint in London, gave great satisfaction and were regarded with full confidence, by both foreigners and natives. For many years, however, this Mint has been entirely under Japanese control, the Imperial Government, according to their invariable custom, having discharged all its European servants after all the information which it thought was required had been obtained.

The buildings at Osaka, begun in the latter part of the year 1869, were rapidly erected. I visited them in October 1870 and found them practically completed, and the necessary machinery was then being installed.

The function of opening the Mint, which took place on the 4th of April, 1871, offered a scene of the greatest enthusiasm, and attracted enormous crowds of people. A very distinguished company of Japanese and European gentlemen came from Tokio and Yokohama to be present at the ceremony.

As will be related on a future page, I had succeeded in obtaining as lighthouse tender a most comfortable steamship, named the *Thabor*, formerly the property of the Messageries Maritimes de France. Placed in charge of a Peninsular and Oriental Company's Captain, with British officers and engineers and a Japanese crew, she was kept in perfect order, having all the appearance of a private yacht. The *Thabor* was at this time probably the only vessel flying the Japanese flag which was at all fit to travel in. She was therefore constantly requisitioned, when she could be spared from her particular service, to convey high dignitaries from place to place.

In this vessel on this occasion there travelled from Tokio to Osaka the premier Sanjo,[3] the Mikado's adviser and then almost omnipotent in the Imperial Government.[2] He had been one of the Nobles of the Court at Kioto, who took a leading part in the great revolution. Date,[4] *daimio* of Sendai, and a member of the great council; Okuma,[5] then Minister of the Interior, an English-speaking officer of the greatest intelligence, under whose authority my Department was at that time placed; Sawa[6] and Kajo, two high Ministers of State, with about twenty others of rank[3] were in the fore cabin. Among the Europeans were Sir Harry Parkes, Mr Cargill, Director of Railways, Mr Morel, Engineer-in-Chief of Railways and Mr W. G. Howell, Editor and Proprietor of *The Japan Mail*,[7] the leading English newspaper published in Japan, who wrote a very able account of the whole affair, with luminous philosophical reflections on feudal Japan.

Leaving Yokohama at 6 p.m. on the 1st of April, 1871, our vessel passed during the night a number of the lighthouses which had, at that time, been established. It was with great pleasure that I seized the opportunity of showing these to the distinguished men on board, who expressed the greatest appreciation of them.

Arriving at Kobe on the morning of the 3rd, a Japanese man-of-war was waiting to convey Sanjo and other Government officials to Osaka, twenty miles distant. These dignitaries preferred, however, to remain on board the *Thabor*, not evincing any desire to patronise their own war vessel.

In two hours Osaka Bay was reached. The forts at the mouth of the river greeted the arrival of the *Thabor* with a salute of twenty-one guns. This was probably the first salute[8] which had ever been fired from Japan's shore.

On the following day the Mint having been formally opened,[4] an immense banquet was given under a marquee, in European style, and toasts were proposed, healths drunk, and speeches made in several languages in the most orthodox manner. Fireworks, some procured from England and some home-made, followed in the evening. The size of the crowds collected to see them and witness the ceremonies, was said to be quite unprecedented in the country.

Having remained at Osaka for some days, the whole party re-embarked on the *Thabor* for the return voyage; during which the vessel stopped at two or three lighthouses, to allow those of the party who desired, to visit and inspect these, eventually arriving at Yokohama in comfort and safety.

CHAPTER TWELVE

THE GREAT FIRE IN TOKIO

ONE OF THOSE disastrous conflagrations, which were of periodical occurrence in old Japan, happened in Tokio on the 3rd of April, 1872.[1] One quarter of the city and many public offices were destroyed.

As has been indicated, Japanese houses simply consist of wooden uprights, supporting a heavy roof and with paper screens between them. They are therefore veritable tinder boxes, and the flames, once getting a good hold, spread with alarming rapidity. The fire-extinguishing appliances then in use were not worthy of the name. Instead of the proper kind of engines, the people trusted in great measure to their gods to put out the fire for them. They brought images of these imaginary deities and set them near the conflagration, feeling assured that its progress would be arrested, and their property be safe.

After this occurence it was desired to erect in the burnt out districts, then a flat bed of ashes, a class of houses not so liable to destruction as formerly. I was approached with a request for a report and plans of what I could advise in the way of prevention and improvement.

Retaining the character of the native house, I proposed a brick shell, tile roof and the isolation of each house by means of a passage between them of at least six feet width. These recommendations which were accompanied by detailed plans were approved and to a large extent adopted. A desire also arose for glass instead of paper windows, and I was requested to obtain men from home who would establish a glass manufactory in Tokio. This enterprise was however indefinitely postponed, and the average native house, excepting that some of them have brick outside walls, remain for the most part the same as they were. In the architecture of public buildings there has been great progress.[12]

CHAPTER THIRTEEN

THE CRAZE FOR STEAMERS

TOGETHER with their adaptability for imitation and their love for novelty, the Japanese of 1870 possessed an extreme desire for toying with European productions, as they would with playthings. In these days of first acquaintanceship with Occidental novelties, the passion for the possession of these arose less from any desire of putting them to practical use, than from the mere idea of ownership. This trait in their character led them at times into many ludicrous situations.

For example, rabbits not being known in Japan, a British trader conceived the idea of introducing a few of these long-eared animals. The result was that the people became completely enamoured of them. An eagerness to possess rabbits at once overran the whole nation.

California, Australia and China were searched for tame specimens. Hundreds arrived by every steamer. The demand for them was nevertheless so great that in their excitement buyers offered as much as $100 for a single rabbit. The Government becoming alarmed took steps to impose a check on the importation, placing a heavy tax on each head.

This had the effect in a short space of cooling the popular ardour, and the sudden passion died out altogether, hardly a single rabbit remaining in the country.

A similar, but much less intense desire arose for the possession of pigs and sheep, neither of which are found in the country, but this mania soon disappeared and gave little trouble. Of other fashionable crazes, from 'the rabbit year' of 1873 to 1889, Professor Chamberlain has written in his unique little cyclopedia, *Things Japanese.*

The first and deepest infatuation of the Japanese of the sixties and seventies, and the most far-reaching in its effects, was for the possession of steamboats.

High officers of the Government, feudal barons, and all who could command sufficient means, purchased steamboats.[1] Now a steamboat is not a toy in an insular country like Japan, but a practical agent of the highest value. Unfortunately for these first purchasers, steamboats are highly intricate, and in the hands of the ignorant, dangerous[1] instruments, both as regards their guidance across the sea, and their internal propelling machinery. Heedless of the fact that their own people were without any experience in controlling or working them, Japanese owners placed unskilled persons in charge of these vessels with boilers in them - frequently with results of a most disastrous character. Since they were too intelligent not to have foreseen the calamities which ensued, the purchase of steamship after steamship by the Japanese can be attributed only to their uncontrollable passion for novelty.

The world was scoured to obtain steamers. The United States, China, India, France and Great Britain all sent their quota - not of serviceable, modern vessels, but of what could be obtained quickly and at a low price. The Japanese wanted something cheap, and they wanted it in a hurry. Worn-out craft, which in many instances had been laid up as useless, were speedily furbished up for sale in this new market. Disdaining the assistance of Europeans and unable of themselves to make any critical examination of hulls or machinery, the Japanese made the most worthless purchases, while not over-scrupulous merchants and agents chuckled.

Inability on some occasions to start, on others to stop the machinery; total collapse of engines; the burning of boilers through want of the needful supply of water, or their occasional bursting with frightful loss of life were usual occurrences.

Vessels on which the men in charge had, as they imagined, learned to handle the machinery, and were thus enabled to proceed to sea, were mostly directed to some rock or sand bank, on which they were either wrecked or had to be assisted off. 'With too many captains, the boat runs aground', ran their old proverb. The 'modern instance', or the new amendment, in practice proved to be - 'With too much motive power, the ship is lost'.

Our lighthouse tender proceeded on six or seven occasions to render assistance to stranded steamers worked by Japanese. On every occasion these craft were found to be in the most filthy and neglected condition. Their machinery was red with rust or bathed in oil and grease. Decks were unwashed and untended, cabins dirty and unsavoury, and ropes and fittings lying about in tangled masses. This was usually the case when foreign novelties were in Japanese hands. It took these people some time to adjust themselves to their new duties of a strange craft. Only after long training did they transfer their habits of neatness to new situations. Then their routine was excellent.

Without charts or sailing directions, the captains of these ships had no knowledge of the rules of the road, and at night the lights were not set. Thus not only were these vessels dangerous to those on board of them, but they were even more so to properly manned ships unfortunate enough to approach them. When one of these steamers broke loose and became helpless, it was like 'running amuck' in port among respectable shipping.

In the case of these wild vessels stranding, no attempts were made by their own crew to save them. Seeming to acquiesce in their misfortune, they remained passive and indifferent,[2] or were perhaps ignorant of what should have been done under the circumstances.

On some occasions, high Government functionaries having been foolish enough to travel in these ships came to grief. When the well-kept English-officered and engineered Lighthouse tender approached, with the object of assisting the Japanese vessel out of her plight, it was ludicrous to observe the stolid dignity, tinged with a slight sense of disgrace, which the rescued man of rank maintained under all circumstances.

Soon learning to estimate justly their own utter incapacity, and as with the growing adoption of European methods, means of travelling by sea became a necessity, foreign officers and engineers were appointed to all the Japanese steamships. Advice was also obtained before new vessels were purchased.

CHAPTER FOURTEEN

LOCATION OF THE LIGHTHOUSES

MY CHIEF BUSINESS in Japan being the erection of lighthouses, the first step was to make an examination of the sites recommended by navigators. As internal means of communication were primitive in the extreme, travelling by sea was the only means of reaching those sites. Hence the provision of a steamship for this special purpose was absolutely necessary. I had decided objections to travelling in a Japanese vessel, even if one had been available.

Sir Harry Parkes, who was most anxious that no time should be lost in beginning the lighthouse work, urged the British Admiral Keppel, on the station, to detail a despatch boat from among the vessels then under his command for this special service. He detached *H.M.S. Manilla*[1] under Captain Johnson, and my first visit round the coast was made in this ship.

I was accompanied on this journey by a Japanese functionary[2] of high rank, who had been a member of the Shogun's council. He had concluded with the foreign representatives the convention of Yedo in 1866, a clause in which provided for the establishment of lighthouses. In the Captain's mess there were also Mr Fujikura, my interpreter, and Mr Blundell, my assistant. In the Ward Room mess were three under officers. In the lower deck, among the blue jackets, were eighteen or twenty Japanese attendants.

Our trip was begun on the 21st of November, 1868. Although bays were entered and places visited which had not been surveyed, or of which no information was available, the voyage was completed without accident on the 5th of January, 1869. The sites for fourteen lighthouses were visited and surveyed, their height above the sea measured; notes were taken of the building materials and labour obtainable at each, and other information procured, all of which was embodied in a report afterwards presented to the Imperial Government and to the British Minister.

As this was the first occasion on which these Japanese gentlemen had sat down to a European meal, their experiences were of a varied character. The curiosity they evinced at the sight of the table, with its cruets, knives and forks and dishes, was exceedingly ridiculous. Entirely without any knowledge of the purpose of the different articles, they put them to all kinds of wrong uses. Catsup and vinegar were severally tasted and wry faces made at them, pepper castors were smelt with distressing results, and beef or mutton looked at askance. For the first few meals the Europeans were kept in uncontrollable fits of laughter, which were much added to by the placid dignity and *sang froid* with which the trials of the unpalatable comestibles were made. However, potatoes and other vegetables found considerable favour.

Nevertheless, as illustrating their extraordinary adaptability to violent change, two or three days only had elapsed before these Japanese seemed to comport themselves at meals as well-bred Europeans would, and to relish every dish set before them. They were not even awkward in the use of their knives and forks, nor did they put the contents of the cruet stand to other than appropriate uses.[1]

In four or five days time the ship reached the small bay of Shimoda, about 100 miles from Yokohama. It was in a temple on shore here that Mr Townsend Harris, the first representative of the United States of America, was located after Commodore M. C. Perry had negotiated the initial treaty with the Shogun. About seven miles off this bay stood a solitary rock,[3] the summit of which was eighty feet above the sea, which all ships from the south bound for Yokohama had to pass. The task of locating the lighthouse on this rock was therefore one of the most important and the most difficult with which I had to deal.

Being many times prevented by wind and sea, and having spent several days either in trying to land on the cone-like rock, or in collecting necessary information, I was informed by his lordship from Tokio that having felt very sea-sick on the voyage from Yokohama, he could not continue it any further but would return overland.

The 'sickness', as afterwards discovered, was a mere subterfuge. The fact was that, while the Tokio Government desired to give the expedition an air of importance by appointing so high an official to accompany it, he was too powerful and useful a functionary to be long absent from his higher duties, and it was prearranged that he was to go a short distance and then return. This practice of deception, sometimes with no apparent object, but at others, with very high aims, was found to be present in most Japanese dealings.

Thus rid of a portion of its burden, the *Manilla* duly arrived at Vries Island,[4] with its perpetual cap of smoke, which is so conspicuous an object on approaching or quitting Yokohama. Its native name is Oshima,[2] (O meaning great, and *shima* island), this being but one of many Japanese islands so named. A splendid harbour of great size is here formed by a large island, whose inside coast line is at a varying distance of half a mile or two miles from the mainland; the intervening water space, being perfectly sheltered from the sea, affords excellent anchorage.

Two lighthouses were required in this neighbourhood, but the headlands being covered with brushwood and trees, considerable time was spent in doing the preliminary work of clearing the space. We were informed by the local authorities that they had received instructions from the Mikado's Government in Tokio to give us every assistance, and I was much impressed by the assiduity with which they carried out my wishes. The daimio of the district sent on board as a present a quantity of sweet potatoes and a number of fowls.

These were acceptable enough, but seeing on shore a number of black oxen used for pack-purposes, an attempt was made to purchase one for food. A price was soon arranged to the mutual satisfaction of both parties. Yet with an instinctive genius, which seemed almost superhuman, these Buddhist folk had divined our purpose in offering to make the purchase. They then resolutely refused to close the bargain. They were willing to sell if the animal died a natural death, but refused if the intention was to kill it. Their compunctions were eventually got over partly by raising the price offered, and partly, it must be said, by the practice of a little harmless deceit.

Calling at several other places on the way, where foreigners had never been before, the *Manilla* duly arrived at Kobe, known in the treaties as the open port of Hiogo. Going from there to Osaka, I carried out my investigations at the mouth of the river, which have already been described.

During one of my interviews with the Governor of Osaka in reference to this work, the latter introduced to me an official of high rank who had specially come from Kioto to superintend the lighthouse work and who would accompany us on our present visit to Yokohama.

At a subsequent interview I was informed by this gentleman, my new chief, that the Government considered it derogatory to the dignity of a person of his rank to travel on official business in a vessel owned by a foreign Government, and that he had determined to purchase one in order to complete the tour of the lighthouse sites; requesting me and my staff to transfer ourselves to it.

It is needless to say that I did not for one second entertain the idea of acceding to such a proposal, conveying as it did an affront to the British Minister, who at great trouble, had obtained the services of *H.M.S. Manilla*. But besides this consideration, I already had sufficient experience of Japanese steamers of that date to make my resolution absolutely inflexible. On seeing the uncompromising attitude which I adopted, my lord without avail used threats of the Government's, and even the Mikado's displeasure, and the possible stoppage of the lighthouse work.

This was the first instance which had presented itself of another characteristic of the higher class Japanese of the old time sort, viz., a haughty and arbitrary demand for tacit obedience, which, while they generally secured this from their own people,[3] was much more difficult to extort from foreigners. Future dealings with Government officers clearly showed me, as will be explained later on, that a definite amount of judicious firmness was necessary, if work was to be accomplished with any success.

Having finished my work at Osaka I rode to Kobe, and took up my old quarters on the *Manilla*. Shortly after my arrival, a steamboat was seen approaching the anchorage with flags flying from every mast and a large Japanese ensign at the quarter. This turned out to be my lord's new purchase. Well known to mariners who were in the habit of visiting Shanghai, she had been employed for very many years as a tug-boat on the Chinese rivers. When worn out, she had been laid up on the bank as having seen her best days. Now, most appropriately named the *Tai-kun*, besides being brilliantly painted and highly decorated, she had a house-like cabin built over her quarter-deck, the floor of this grand apartment being spread with *tatami*, or thick Japanese mats. Sent to Osaka, this hopeful craft

70

was, without advice or examination, bought by my lord for the lighthouse service.

I lost no time in informing my Japanese friend of the character of his purchase, and was not surprised to learn soon after that he had abandoned her. Nevertheless, determined not to commit the imaginary indignity of travelling in a British ship, he had chartered another old steamer, the *Argus*, belonging to a firm of British merchants in Kobe. As I still refused to join him in this, he expressed his determination to accompany the *Manilla* in it. This he did for two days after leaving Kobe, the two vessels steaming slowly along the Inland Sea, stopping for our visits to certain points on the way. Eventually we reached Hiroshima.

Here an incident occurred, which is worth narrating, as showing another, but this time a most pleasing and praiseworthy phase of Japanese character. One of the midshipmen on the *Manilla*, a lad of nineteen, died quite suddenly while the vessel was at anchor. He was buried on the shore of the beautiful bay, the whole ship's crew accompanying the coffin. The officers, fellow-countrymen of the dead youth, stood in a group at a respectful distance, while the ceremony was proceeding. When it was over they approached, and the Tokio magnate made a sympathetic little speech. He finished it by saying that in Japan it was the custom to present flowers to the dead; but as there were none in this locality, he asked permission to place a headstone at the grave, and had written to his Government asking that orders should be given to have the tomb carefully preserved and taken care of by the local authorities. Immediately after the funeral ceremony was over, a very pretty sight was presented by quite a number of aged men and women approaching with shrubs and twigs, which they reverently laid on the grave. These proceedings at Hiroshima considerably enhanced the Europeans' opinions of Japanese character, so far at least as kindliness of disposition is concerned. I visited the grave some years afterwards and can testify to the fact of these instructions being, up to that time, carefully carried out.

Proceeding from Hiroshima towards the Straits of Shimonoseki, we lost sight of the *Argus* after dark. Arriving off the town of Shimonoseki in the early morning, I visited the Governor alone, the *Argus* not having arrived.

CHAPTER FIFTEEN

IN THE HISTORIC PORT OF NAGASAKI[1]

HAVING COMPLETED all the needful 'business' and being within two days of Christmas, at Captain Johnson's urgent request, he being desirous that his crew should have that day free, I decided to await no longer the arrival of the *Argus*, but to proceed to the important treaty port of Nagasaki,[1] where the *Manilla* duly arrived on the 24th of December.

Next day the *Argus* arrived, but without any Japanese. She had run ashore somewhere in the neighbourhood of Shimonoseki, and after some trouble was re-floated, but my lord and his troop of retainers, becoming somewhat distrustful of their own personal safety, abandoned their vessel, and set out overland to Nagasaki.

The Governor of Nagasaki, Inouye,[2] I discovered to be a young man who had been in the United States, and was partly educated there. As he spoke English fluently, I carried on all the business that was necessary, having in all my proposals his willing and intelligent cooperation. He was intensely amused at my relation to him of the high officer's escapades with the *Tai-kun* and *Argus*. He expressed the belief that 'being a good Japanese, he would soon gain more sense'.

In four or five days time, the official traveller overland appeared. While he seemed to waver somewhat as to accepting a passage in the *Manilla* for the remainder of the voyage, he came to the conclusion that since two had failed him, and he was still only about one hundred leagues from his starting point, he would purchase or hire no more steamers.

After due consideration, he finally decided to abandon the whole trip and betook himself and his followers back to Osaka overland.

Such was the typical functionary of old Japan. Having, from a petty feeling of pique, refused to use the vessel generously placed at his disposal by the good offices of the British Envoy, and having bought one vessel and chartered another, he finally returned to his home in a palanquin, with his large retinue following on foot. He had accomplished not one atom of useful work, nor had he in any way advanced the construction of lighthouses.

I feared that the not inconsiderable sums which he must have spent would be debited against the cost of the lighthouses, so I sought to obtain information as to the amount of his expenses. This has always been refused me, and to this day remains a profound secret. Such an official, having been entrusted with the inauguration of lighthouse work in Japan, gave me little hopes of its smooth or successful accomplishment. In my eyes, no man could have presented a more unhealthy aspect of ignorant bumptiousness, dignified stupidity, and unreasonable stubbornness than he did.

Nagasaki, a specially interesting port with a harbour entirely land-locked by lofty hills, is one of the most beautiful, even as it is the safest in the world. The foreign settlement, as well as the native town, are picturesquely situated on the face of the hill on one of its sides, as the many thousand Europeans know well who have visited the place. Many works of importance were being carried on at Nagasaki. A patent slip for good-sized ships had been laid down and a series of workshops containing a quantity of tools established, where Japanese alone were at work, and seemingly fully employed. An English firm of merchants, the Messrs Glover,[32] had succeeded in making an arrangement with the Government in Tokio for the working of coal found on the neighbouring island of Takashima.[4] To my knowledge, this was the only instance of any European being allowed to mine this mineral in Japan.

Being once present in Tokio at an interview between Sir Harry Parkes and Terashima, then the Minister of Foreign Affairs, I heard the former try to impress the latter with the great advantages likely to accrue, were the Government of Japan to allow its great mineral resources to be utilised. The reply of Terashima was that the minerals of Japan belonged to the Japanese, that there was no fear of their deteriorating through keeping, and that they must remain unoperated on until the Japanese themselves were in a position to work them.

I visited the workings at Takashima and found that three hundred men were daily employed in them, that modern hoisting plant and pumping machinery had been erected, and that about two hundred tons of a good bituminous coal were being raised daily. The fuel found a ready sale at $4.50 per ton, English imported coal being $7.50 per ton.

On the point recommended for a lighthouse near Nagasaki, it appeared that an iron tower, made at the foundry on shore, had already been erected. In this, a curious instance of perverted ingenuity was displayed. While the tower itself was sufficiently strong and large, a flimsy lantern had been placed on its summit, in the centre of which was placed an ordinary flat-wick paraffin oil lamp, without lenses, reflectors, or the other usual means of conveying light to a distance. In order to accomplish this efficiently, it may be unnecessary to inform readers that the media through which the light passes should be as clear and bright as possible, in order to prevent the absorption of rays. While the small, flat-wick lamp was in any case an inefficient illuminant for piercing to any but the smallest distance, it was rendered completely useless by the panes of the lantern having had pasted over them sheets of tissue paper. Enquiring of those in charge the purpose of thus obscuring the lantern, I was somewhat taken aback by their reply. It having been observed, said they, that in foreigners' houses in Nagasaki the lamps have shades of ground glass around them, the conclusion come to was that these were used in order to increase the power of the light. They had tried, but were unable to obtain any ground glass, so they used paper, as they considered it to be a suitable substitute. Our own want of scientific method had led the Japanese into a whimsical error.

The Japanese, however, affected to believe that this lighthouse, as it stood, was perfectly satisfactory. They informed me that captains of British ships had expressed themselves as pleased with it, that $7,000 had already been expended on it, and that they objected to anything further being done.

The real facts of the case were that, urged by foreign merchants to erect a lighthouse, the Japanese with their natural self-sufficiency proceeded to do so in their own way, and their production was as has been described.

Feeling sure that the statement that British captains of warships had reported in its favour was not accurate, I contented myself by informing the Japanese at Nagasaki of the stipulations in the treaties, which would necessarily have to be carried out, and that they would probably receive instructions from Tokio regarding them.

One very important point for a lighthouse still remained to be examined, viz., what was at then the most southern extremity of Japan and our course was laid for Satanomisaki. On arrival it was found impossible to land owing to the high sea. No sheltered bay in the locality being known to the captain, it was determined to postpone examination until another occasion, and to return to Yokohama, which was safely reached on the 5th of January, 1869.

CHAPTER SIXTEEN

BUYING A LIGHTHOUSE TENDER

THIS PRELIMINARY trip over, and arrangements for furthering the work having been completed, the necessity for a steamship for the lighthouse service became apparent. I was instructed to report to the Government in Tokio, whenever the opportunity of purchasing one properly fitted for the purpose should arise.

Soon after this a new but small screw steamer, which had come from England with cargo, named the *Sunrise* hove in port. Of between 400 and 500 tons burthen, she had two good-sized cabins and the necessary sleeping accommodation. Having had her boilers, machinery and hull thoroughly examined, and favourably reported on, by engineers from British men-of-war then at Yokohama, and Sir Harry Parkes expressing his approval, I laid before the Japanese Government a proposal for her purchase. Her captain, being owner, had placed her sale in the hands of agents, and through them the price to be paid for her was fixed at $60,000, to which the Japanese assented. One quarter of the purchase money was to be paid on taking possession, when the change of flag would be made, the other three quarters at intervals of one month for each.

My Japanese superior officer had, by the time that these arrangements were completed, finished his overland journey from Nagasaki, and he again took up his somewhat erratic and always obstructive direction of lighthouse affairs.

On the 24th of January, 1869, the first instalment of $15,000, in bullion, arrived from Tokio, and was duly delivered to the agents. Then the Japanese flag was hoisted on the *Sunrise*.

On the very same day on which this transaction was completed, a direction from my lord was received by me that the vessel was to proceed to Osaka on other Government business.

The *Sunrise* had been purchased for a definite purpose, in which all her services would be required, and it was sought at once to divert her from that! Feeling that if this were allowed, her usefulness to the lighthouse service might be wholly nullified, I protested as strongly as was in my power against such an action. For three hours did I wrangle with my lord, but to no effect. That individual would not be moved. I then referred the matter to Sir Harry Parkes, who agreed with me completely as to the danger of permitting such procedure, and promised to see the higher officers in Tokio concerning it.

A matter of another nature, however, prevented the bureaucrats' wish being carried out. The captain and owner, who had received but one instalment of $15,000, objected to the vessel's sailing until further sums were paid. As the old officers and crew were still on board, the Japanese could do nothing by themselves. The *Sunrise*, therefore, did not leave her

77

moorings at this time, and, although the Government paymasters hurried down from Tokio with the remainder of the purchase money, so that in about a week a clean receipt was given for the whole, the use of this ship for other than lighthouse work was not persisted in.

Struck by the intelligent demeanour of Captain Brown,[1] one of the chief officers of one of the Peninsular & Oriental mail steamers, in which I had travelled to Japan, I sounded him as to whether he would resign his position in that company and take service with the Japanese Government as captain of the lighthouse tender. Expressing his willingness to do so, and terms having been arranged, he was on the 24th of February, 1869 placed in charge of the *Sunrise*. He engaged English officers and engineers, with a native crew.

Proving himself to be a well-educated, willing, very obliging and extremely cautious man, he became a favourite with the Japanese. As has already been stated, he assisted in making a survey of the Bay of Yokohama. He also surveyed many of the unknown harbours into which his attendance on the lighthouses obliged him to go.

After Captain Brown's appointment to the *Sunrise*, having made several short trips to lighthouses in the immediate neighbourhood of Yokohama, a voyage round the whole southern coast was commenced on the 7th of July, 1869. The Japanese officer who on this occasion accompanied me was Inouye. Speaking English with ease and correctness, he seemed to have imbibed all the sense of humour and restless energy of the American people, among whom he had been educated. No Japanese I ever encountered had the same flow of spirits or animation. He poured out undiluted ridicule on the antiquated methods of his compatriots, and startled them by the vigour of his methods. None of my voyages along the coasts of Japan was so pleasant and free from childish disputes with petty officials as

Inouye rendered this one.

No department in the Government service made such good progress as did the Lighthouse Service, while Inouye was connected with it. Too valuable a servant of the Emperor to remain long in a subordinate position, Inouye was transferred and received rapid promotion in other departments of the Government. In 1878, he became Minister for Japan at the Court of St James. As such, he kindly consented to be present at a meeting of the Institution of Civil Engineers in London in 1878 when I read before that body a paper describing the lighthouses of Japan.[21]

I was accompanied also by my wife and Colonel Brasier, a retired Indian officer then on a visit to Japan, while inspecting the positions at which lighthouses were being erected.

CHAPTER SEVENTEEN

MY VISIT TO SATSUMA

ONE OF MY most interesting visits was paid to Kagoshima, the capital city and headquarters of one of the most famous of Japanese clans, and to other points in Satsuma. Kagoshima was bombarded by the British fleet in September 1863 as a reprisal for an attack[1] on a party of four English persons by the retainers of the Prince on the highroad near Yokohama. As a result of this bombardment, a just appreciation of Western methods was developed among the Satsuma men, and many evidences of this were already visible at the time of my visit.[1]

The lord of Satsuma,[2] powerful and to a great extent independent, was still, in 1870, regarded with great distrust by the Imperial Government. In fact, Inouye, not feeling assured in his own mind as to what the nature of our reception would be, requested that none of our party should go on shore till he himself had sounded the disposition of the local authorities. Leaving the vessel soon after her arrival, he returned in four or five hours' time, accompanied by four officers of the ken, or prefecture. These welcomed us in the heartiest way, and expressed their desire to do anything in their power to assist us.

They informed us that the lord of Satsuma, usually called 'the Prince', desired us to dine with him on the following Tuesday evening, and hoped we would accept the invitation. Having asked them to convey to the Prince our thanks for his condescension, I expressed the pleasure which all would feel in sharing his hospitality. They then informed me that they were extremely sorry to say that the Prince had no wine, and asked if they could have some from the ship. Six bottles of champagne and six bottles of sherry were our contribution to the feast.

On the following morning, the local officers again came off to the ship and received explanations regarding the work proposed to be done at the lighthouse in Satsuma. The point selected was Satanomisaki, forty miles from Kagoshima, which I was unable to land at during my voyage in the *Manilla*.

The officers took our party on shore and showed us a huge cotton factory,[3] which was in seemingly perfect order and at work to its full capacity. The machinery used had been made in Oldham, England, and was imported through Messrs Glover of Nagasaki. Our party was also shown the Arsenal, where good-sized cannon and ammunition were being made. Various other useful trades were in prosecution, such as boat-making, glass-blowing, carpentry, etc. A guide was instructed to take us through the town, but for some reason he would not allow us to visit the most interesting places, such as the Prince's 'Palace', or the temples, but confined us to dirty narrow streets. The thermometer being at the time over ninety degrees in the shade, we were glad to leave these uninteresting and tiresome quarters.

The evening on which we were invited to dine with the Prince having arrived, we were conducted to a large room within the cotton factory. A long table was laid with about twenty covers, the ceramic ware[4] having been made in the Staffordshire potteries, and the orthodox cutlery in Sheffield. The Prince himself, it was explained, was not able to be present, but he had sent his chief officers as his representatives. Furthermore, he desired our acceptance of some presents from him. These consisted of some exquisite pieces of Satsuma faience and silk. These souvenirs of hospitality I still cherish, Satsuma being the most noted of all the provinces of Japan for the production of the first named articles. The dinner was a well-cooked and well-served repast in Parisian style, which was seemingly much relished by the Japanese.

The striking oddity was that after every course, from soup to pastry, had been partaken of, and the meal presumably finished, a beginning was again made. Soup was once more brought on, and the other courses followed as before. Having gone through these a second time, it was evidently the host's intention to present them once more; but I, having hinted to Inouye that this was unnecessary, and he having conveyed the hint to the Satsuma officials, matters were brought to a stop and the dinner ended. The healths of the queen, Victoria, The Mikado, Mutsuhito and the Prince, Shimadzu Saburo, were duly honoured, after the orthodox European fashion, but conversation rather flagged from want of interpreters.

This dinner afforded a further illustration of the

Japanese love of innovation, because being far from any European settlement they had no opportunity of studying European ways. Their most ornate meals[5] are simplicity itself compared with that which they gave to the European party. It is probable that some of the Prince's household had gone to Shanghai or some other open port in China, and there learned European methods of cooking and eating, instructing others on their return; but I have never learned how they obtained their information.

In some return for their general kindness, Captain Brown and the chief engineer of the *Sunrise*, along with myself, at their request inspected a small steamer which the Arsenal officials had been attempting to construct, but over which they had got into difficulties. Such advice and assistance as was in our power we gave cheerfully. Expressing themselves as very thankful for these, they put much too high a value on the aid afforded them, presumably from an excess of politeness.

A solitary Englishman, Dr W. Willis,[62] was then living in Kagoshima. He was a man well known in Japan and much respected by both foreigners and Japanese. For many years employed by the British Legation, and being an exceedingly skilful surgeon, he gave invaluable assistance during the troubled times Japan had just passed through. With Mr (now Sir) Ernest Satow, then student interpreter, in the early part of 1868, he was the first foreigner to enter the city of Kioto. He had been specially asked by the Mikado's advisers in the new Imperial Government to come and succour the wounded and sick after the fighting which took place, especially at the battle of Fushimi. The lord of Satsuma was most appreciative of Dr Willis' services at this time, and pressed upon him a high pecuniary reward. This, with much firmness, he refused absolutely.

When matters had settled down, however, the

Prince invited Dr Willis to take up his residence in his capital, and give himself and his people the benefit of his medical advice. Having consented to this at a reasonable rate of remuneration, for a period of five years, Dr Willis was undergoing, when I met him, this term of voluntary banishment.

Right glad to hear the voices of his own compatriots once again, Dr Willis almost lived on board the vessel, while she remained at Kagoshima. He described the people amongst whom he lived as inoffensive and kindly in their ways, but devoured by the most degrading lusts, and, in consequence, suffering from the most loathesome diseases.

Having little or no means of relaxation or enjoyment, Dr Willis devoted his time partially to a study of the Japanese language, and partly to reading the Encyclopedia Brittanica, the whole twenty-eight volumes of which he had in his library, and every line of which, he informed me, he was determined to read and digest.

On departing from Kagoshima, I proposed visiting the site for the lighthouse at Satanomisaki, and the Satsuma officials announced their intention of accompanying the *Sunrise* in one of their own steamers. Leaving at the same time, the *Sunrise* accomplished the forty-mile run in two hours less time than the Satsuma steamer, during which I had almost completed the observations I found necessary. I found the site to be a most difficult one, being on a pinnacle-shaped rock separated from the mainland by a dangerous channel of some hundreds of yards width, through which the rapid ocean currents and the unbroken Pacific Ocean rollers constantly drove and leapt. The top of this lonely island was no larger than was necessary to take the foundations of the lighthouse.[3]

The Satsuma officers on their arrival made the attempt to land on the rock to join me, but failed. In making the attempt, their boat was in great risk of being swamped, while one or two of them were in such a state of collapse from seasickness, that they were incapable of movement.

Returning to their steamer, I joined them, and explained to them what my probable intentions would be with regard to the lighthouse. On bidding them good-bye, and thanking them for their attention, they expressed a desire that the *Sunrise* should accompany them half-way back to Kagoshima, as they were unable to manage their own ship and feared an accident. The real fact was that darkness was coming on, and if they lost sight of land, or possibly had no compass on board, their vessel might founder. To their request Captain Brown was strongly opposed, and Inouye and I had to agree with our commander, so the Satsuma men were left to get back as best they could. This incident afforded another instance of the utter untrustworthiness in nautical matters of the Japanese of that day.

CHAPTER EIGHTEEN

THE AMERICAN WARSHIP
ONEIDA

AFTER COMPLETING the voyage to Satsuma, the *Sunrise* did good work in towing disabled Japanese steamers into port or assisting them off rocks and sand-banks. She made many trips in the service of the lighthouses, proving herself to be a good and efficient vessel.

One especially mournful service did she perform. On the 25th of January, 1870, the Peninsular and Oriental Mail Steamer *Bombay*, was entering the bay at Yedo after dark. When about twelve miles distant from Yokohama she met the U.S. man-of-war *Oneida*, which had started two hours previously on her way to San Francisco, taking home time-relieved men of her own and from other ships on the station. The two vessels came into collision, the *Bombay* striking the *Oneida's* bow in mid ships, cutting her almost in two, and causing her to sink in a few minutes. One hundred and ten of the officers and men on board the American ship were drowned.[1] The *Oneida* being a wooden ship, her timbers pierced the iron bow of the *Bombay*, becoming jammed in the holes, and were taken away by her in that position.

On whom the blame for this lamentable accident was eventually thrown I have no record. It had been asserted by one of the survivors, that he had heard Captain Eyre of the *Bombay* make use of the expression 'Served them right', or something to that effect. This aroused extreme indignation among the American population in Yokohama, which rose to such a pitch that a guard had to be obtained from the British men-of-war in the harbour to protect Captain Eyre from their alleged threatened violence.

The vessel being sunk in the fairway of the channel to Yokohama, I was instructed to find the wreck, and if necessary, to place a buoy over it. Accomplishing this in the *Sunrise* after great trouble, and with the melancholy accompaniment of picking up dead bodies, it was discovered that the hulk of the *Oneida*[1] lay in a depth of water of twenty-five fathoms; and, as no vessel could touch it, no buoy was necessary.

In 1870 the extent of lighthouse work became much enlarged. Many of the lighthouses were nearing completion, and as a number of additional ones had been ordered, it soon became evident that the *Sunrise* did not afford sufficient accommodation either for the material or for the men whom it was necessary to convey to the different points. The Telegraph Department, which had been making rapid strides, found itself also in need of a steamer to convey its men and material to different parts of the coast, and it was considered that the *Sunrise* would answer the point admirably. It was therefore decided to transfer the *Sunrise* to this Department, and to obtain a larger and more powerful vessel for the lighthouse service.

CHAPTER NINETEEN

THE PURCHASE OF THE *THABOR*

ABOUT THIS TIME there arrived in Yokohama the Messageries Maritimes Steamship *Thabor*, a powerful vessel of about 750 tons. Honoured as having carried the Empress Eugenie[1] on the occasion of the opening of the Suez Canal, she has already been mentioned as taking a distinguished company to the opening of the Mint at Osaka. Being precisely the character of boat which was required for the purpose of the lighthouse service, I made inquiries as to the possibility of purchase. The Messageries Maritimes Company were willing to sell, and fixed her price at 500,000 francs, or £20,000. I then approached the Government officials, who resolutely refused to move in the matter.

Sir Harry Parkes becoming interested in the question, appreciating the necessity of something being done, and losing no opportunity to improve the interests of foreign commerce to the advantage of the

1. LIGHTHOUSE DEPT.   2. NIHON ŌDŌRI (NIHON BLVD)
3. YOKOHAMA PARK   4. YOSHIDA BASHI (YOSHIDA BRIDGE)

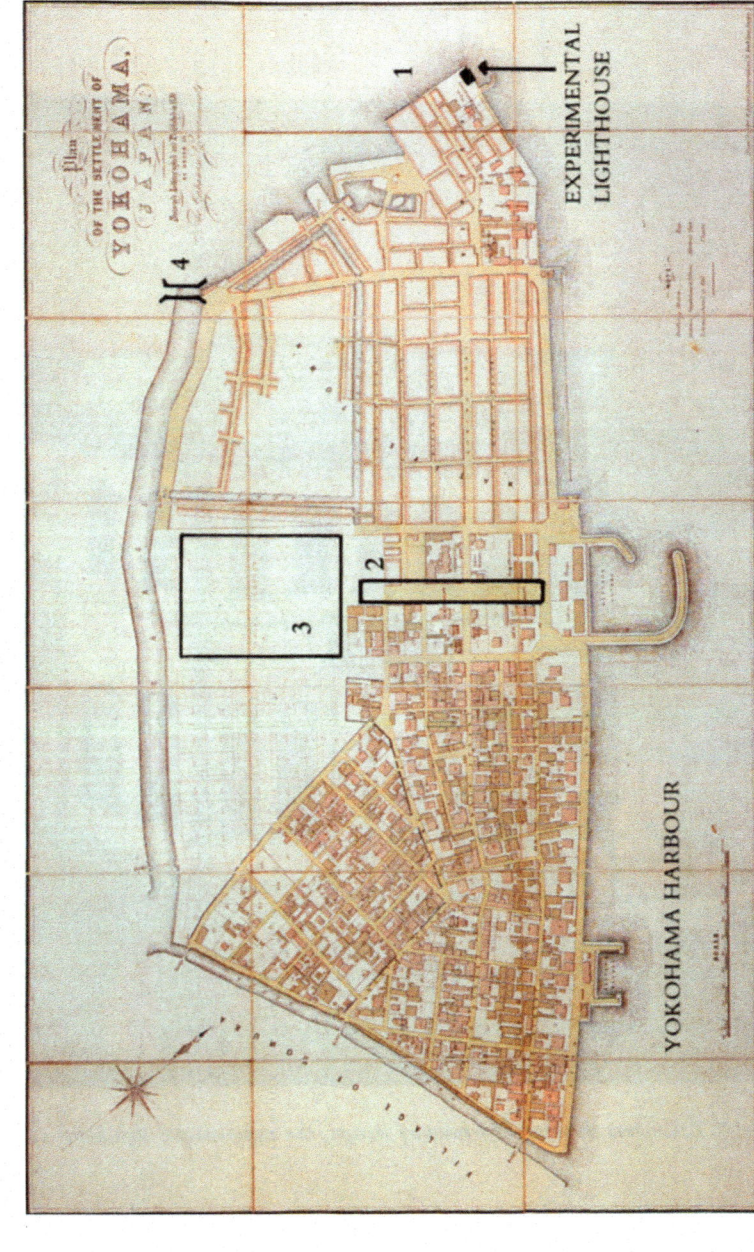

pl.1 *Plan of the Settlement of Yokohama prepared by R H Brunton (1870)*

pl.2 *Print of Yokohama showing first railway station, the experimental lighthouse (top right), the*

aten Bridge (below the lighthouse) and the Ōe Bridge (bottom right)

pl.3 *1870 print celebrating the Yoshida bridge*

pl.4 *Early photograph of Brunton's iron lattice bridge (Yoshida bridge), 1869*

.5 *Nihon Ōdōri (Nihon Boulvard), Yokohama, designed by Brunton in the 1870s*

pl.6 *Brunton photographed in London c.1868*

pl.7 *The Lighthouse Department, Yokohama*

pl.8 *Brunton's residence in tne grounds of the Yokohama Lighthouse Department*

pl.9 *Kashinosaki lighthouse, completed August 1870*

pl.10 *Mikomoto lighthouse, completed January 1871*

Photographs of the Brunton lighthouses courtesy Tokokai —
Japan Lighthouse Association

pl.11 *Irōzaki lighthouse, completed October 1871*

pl.12 *Tsurugisaki lighthouse, completed March 1871*

pl.13 *Yesaki lighthouse, completed May 1871*

pl.14 *Tsurushima lighthouse, completed July 1873*

pl.15 *Wadanomisaki lighthouse, completed May 1871*

pl.16 *Sugashima lighthouse, completed July 1873*

pl.17 *Inubōsaki lighthouse, completed November 1874*

pl.18 *Hesaki lighthouse, completed April 1872*

l.19 *Tsunoshima (Brunton ref. 'Kadoshima')*
ghthouse, commenced August 1873

pl.20 *Noshiap lighthouse, completed August 1872*

pl.21 *Shirasu lighthouse, completed March 1872*

pl.22 *Iwoshima lighthouse, completed January 1872*

pl.23 *Mutsure lighthouse, completed December 1871*

pl.24 *Yeboshishima lighthouse, completed August 1875*

pl.25 *Satanomisaki lighthouse, completed December 1871*

pl.26 *Nebeshima lighthouse, completed February 1873*

pl.27 *Anorisaki (Brunton ref. 'Matoya') lighthouse, completed September 1872*

pl.28 *The lighthouse tender* Thabor, *'a powerful vessel of about 750 tons'*

Japanese, used his influence in Tokio to good purpose. After the lapse of a few days, I was instructed to have the vessel examined and her machinery and boilers inspected by experts. This was accomplished by the aid of the engineers of H.M. ships, and favourable reports were received. Nevertheless, much hesitation was shown by the Japanese in authority to sanction the purchase. It is believed that this was caused chiefly by a shortness of cash in the Treasury, as stipulations were made to pay by instalments, and the whole sum in not less than nine months. To the British minister they made the excuse that I was pressing her purchase on them vigorously and that I was seeking to have her entirely in my own hands. In reference to this latter, I told Sir Harry Parkes that I should acquiesce in any arrangement he should consider best for the service.

On the 26th of March, 1870, I received written instructions from the Imperial Government to conclude her purchase. I wrote out an agreement and obtained the signatures of both parties. Before taking the vessel over, a trial trip was made on the 4th of April, but no Japanese officer made his appearance. The test having proved satisfactory, the *Thabor* was handed over to her new owners, and the Japanese flag hoisted at her peak. Placed in charge of Captain Brown, she became the lighthouse tender, and was soon much sought after by persons in office desirous of travelling along the coast.

The purchase of this steamer was the cause of considerable personal annoyance, revealing social conditions at the treaty ports. There were in Yokohama, as it may be assumed there are in other places, middle men or 'go-betweens', as the Japanese say, who act as commercial agents between aliens and natives. These gentlemen, chiefly Englishmen, are called brokers, and make their living by placing themselves between two parties desirous of concluding a trans-

87

action on which commission may be charged. Partly through ignorance and partly because I did not see that their assistance was required, I committed the grievous sin of arranging this purchase for the Japanese Government without employing one of these curiosities of civilisation. Incurring their extreme displeasure by so doing, I was for some time an object of their malignity. Without being able to trace them to any particular tongue, I heard that reports were in circulation that I did not employ a broker because I myself wanted the broker's fee, that I was in receipt of a commission of ten per cent on the purchase money, that I gave for the steamer whatever was asked and that she could have been purchased for $10,000 or $20,000 less, had the sale been 'properly' negotiated, etc. etc.

As time went on these scandals spread and increased in magnitude. They culminated in a visit which was paid at my house at four o'clock one morning, by a busy-body, known in Yokohama as 'public-spirited' Jack. He was secretary to the Club, a very important institution in Yokohama, and one in which most of the scandal floating about the settlement was manufactured.

This personage informed me that an editor of one of the English papers had received a letter from a good source, saying that I had accepted a commission of $10,000 for the sale of the *Thabor*, that the Government had decided to summarily discharge me, and that the engineer appointed for the railway was to take my place. I proceeded at once to Sir Harry Parkes and told him of this visit. He advised me to pay no heed to, and take no action regarding such slanders, they most likely being only talk. He added that my position was very different from that of Mr M., that I was selected by the British Government to carry out certain treaty stipulations, and before I could be superseded, the Queen's Government would have to be consulted.

I laid the scandals also before the Japanese Government officers, and they expressed their perfect satisfaction with the way in which the purchase of the steamer had been made, and their utter unbelief in the stories which had been bruited about.

Having applied for and received a letter from the agent to the French Mail Company, declaring the perfect honesty with which the whole transaction had been carried out, I sent copies of this to the British Envoy and the Imperial Government and the whole affair rapidly dropped out of mind.[1]

CHAPTER TWENTY

THE JEALOUS JAPANESE

FOLLOWING the practice of European countries, I suggested that the *Thabor* should carry a pennant, thereby designating that she belonged to the Imperial Japanese Government and was engaged on national service. To this no objections were raised, and shortly after her purchase her own distinguishing flag was flown from her mast head.

Towards the end of the year, however, one or two men-of-war having been procured by the Japanese Navy Department, officers in charge of these raised objections, and requested that any display of this particular flag should be discontinued. Their contention was based on the ground that the *Thabor*, being entirely under foreign officers, and, to a great extent, under the management of outsiders, could not therefore be regarded as entirely a Japanese national vessel.

This somewhat whimsical idea shows the jealous bent of the Japanese mind even at this period. That it had influence with the Japanese Navy Department was evidenced by the fact of a peremptory order coming from Tokio, in February 1871, ordering the *Thabor's* pennant to be hauled down. This was followed later on by the following notification showing that the question of flags was much exercising the Government mind at the time.

Copy of Government Notification:-
No. 371

TRANSLATION NO. 63 B

'The undersigned hereby notifies that, in order to prevent an irregular manner in decorating vessels with Flags for Universal Salutation, or hoisting, or hauling our national Flag according to the length of daytime, all the vessels of European style belonging to our Government and merchants when they are in Port will in future perform these by instructions of a ship of war, which is in the same port.'

(Signed) Iwakura.[11] Udai-jin

Strictly adhered to, this notification became somewhat obnoxious to Europeans or Americans in charge of Japanese craft, preventing as it did, the decoration of the vessels on any of the holidays which it had been their custom to honour. On New Year's Day, 1875, the then Captain of the *Thabor* had the audacity to hoist a few flags, which act was immediately followed by an indignant expostulation from the Captain of the Japanese War Vessel in port.

As showing the native fierceness of the Japanese naval officer of that day, this gentleman, deeming the Captain of the *Thabor* did not show the proper amount of contrition for his disregard of the Tokio Government's decree, hurled at him the most unseemly threats and displayed all the rancour of the oriental despot. It

was not until, assisted by the Japanese lighthouse officers, I had interviewed him, and by expressions of
regret, had soothed his ruffled dignity, that he agreed
to allow the matter to drop. What action he had the
power or authority to take was never clearly demonstrated, but his denunciations were of the most violent
character.

CHAPTER TWENTY-ONE

THE DOCKYARD AT YOKOSUKA

THE *Thabor* had done good service for three years,
and so it was thought advisable to give her new boilers.
How to get this work accomplished was a matter of
great perplexity. The opinion was largely held, even
by the Japanese themselves, that sending her to Europe
for the purpose would involve less risk and expense
than attempting to get the work done in Japan. It was
eventually decided to place the matter in the hands of
the French servants of the Japanese Government, who
had built and equipped an arsenal at Yokosuka about
ten miles south of Yokohama.[11]

Before the Shogun's Government had been driven
from power, a large staff of French Engineers had
been obtained and furnished with complete and elaborate machinery necessary for the establishment of a
naval arsenal.

Arriving in 1867, these Frenchmen erected the
machinery, commenced the formation of a dry dock
and made other preparations for the building or repairing of vessels of a large size. This dock, three hundred

91

feet long, with twenty feet of water over its sill, was opened on the 26th of March, 1871. By that time, also, the various machine shops having been put into working order, the Mikado paid the dockyard a visit of inspection on the 1st of January, 1872.

The arsenal at Yokosuka was thus in possession of all the material necessary for the execution of refitting the *Thabor*, but the Japanese workmen were unequal to the task, and the French artisans being so few in number, little reliance could be placed upon a promise of prompt completion of any piece of work.

It was therefore with much misgiving that the *Thabor* was handed over to those in charge of the Yokosuka dockyard.

These French gentlemen received instructions to undertake the work in April 1872. They at once decided to order the boilers to be made in France and sent out orders. The task before them, therefore, was only that of taking out the old boilers, replacing them by the new ones, and renewing any faulty parts in the ship's hull.

The new boilers having arrived in August 1873, the *Thabor* went to Yokosuka. By October, practically nothing had been done to the ship. Feeling anxious, I offered to obtain competent men, and send them to complete the work, but the French Superintendent would not hear of this. On the 7th of February, 1874, some progress had been made, but it was reported that three days' work had, at that time, not been done during the foregoing three weeks. On the 28th of May the French gentleman in charge declared to me that he could give no date for the completion of repairs. On the 6th of July, or nearly one year after her arrival, the *Thabor's* new boilers were in place and connected with the engines, but the decks over them had not been laid, nor had the other finishing woodwork been completed.

Meanwhile, having great cause for anxiety regarding the lighthouses, which had for so long a period been left unvisited, I brought the ship to Yokohama, placed in her a mob of Japanese carpenters, hired a number of Chinamen to do the necessary painting, and by day and night work, got her ready for sea in a few weeks.

This complete indifference to time and to the exigencies of circumstances, which at this period was so characteristic of the Japanese in all their dealings, did not seem to be much interfered with by their French *Yatoi*, or 'hired' foreigners. Having spent a year in taking out old sets of boilers and putting in the new, properly but a few weeks' work, they placidly submitted to the completion of the task, properly theirs, being brought to completion under my direction.

The *Thabor*, already by universal consent a most suitable vessel for the work she had to do, became virtually a new ship, capable of many years work.

Notwithstanding these facts and the large sums spent on her, the Imperial officers, during my absence on leave, and without consulting me in any way, ordered to be built by Messrs Napier of Glasgow, another vessel for the lighthouse service. She was a twin-screw steamer*of about the same size as the *Thabor*, but not nearly so suitable for the work. The inducement which led these men, all drawing salaries from the public purse, to order this vessel, is absolutely unknown to me, but there can be no doubt that their doing so was a wanton and perfectly unjustifiable waste of public funds. Arriving from Scotland this costly steamer became a white elephant, not required for lighthouse work and unsuited for ordinary purposes.

The following description of the Japanese official in those days given by a man who was in a position to be thoroughly conversant with the facts, may afford some explanation of their action.

'There is a system of wide corruption among the

officials in different parts of the Empire. It would seem that the perversion of public money is by many hardly looked upon as discreditable; and from the miners, for instance, whose hours of work were put down in the accounts at double the real number, in order to defraud the local Government, up to the highest officials, the existence of a wholesale system of plunder is only too evident.'

So disgusted was the British Minister with this questionable transaction, that he refused, on many occasions, the invitation from high officers of the Japanese Government, to travel in the new steamer, and accepted such only on condition that the *Thabor* was provided for him.

After I had left the Japanese service, the Government sold this vessel to some native merchants, and the unfortunate craft, meeting the fate of nearly all Japanese steamers, which in the seventies were navigated by natives, was run ashore and totally wrecked.

Recurring for a second to the corrupt character of official transactions, it may be mentioned that when the steamboats reached Nagasaki, it was the custom to fill them up with coal, the city being in a district producing this fuel. The material was generally obtained from some Japanese merchants, who had brought it from a mining district, forty miles to the north, worked by Japanese. On one occasion I was startled to learn that the official accompanying the vessel, who was deputed to purchase this, had been charging for it, in the accounts he rendered, $6.50 per ton, whereas the actual price as I learned from the merchants, was not more than $4.00. Assuming that 5,000 tons per annum were purchased, an amount of $12,500 each year would become the perquisite of the plunderers from this item alone.

While deeming it futile to make any definite charge of peculation, I thought it my duty to inform the coal merchants that the Tokio Government was being made to pay much more than was necessary for coal, and to advise a reduced charge. The usual ingenious excuses were, of course, made, all of which were too apparently absurd to deserve consideration. Promises were given that greater care would be exercised in future, but whether these were fulfilled or not, I am not in a position to say.

*According to Yokohama City archives, the vessel referred to here by Brunton was the *Meiji Maru* which was actually put to extensive use over many years, by no means 'a white elephant'.

CHAPTER TWENTY-TWO

THE EXPEDITION TO FORMOSA

AT THIS TIME the Japanese Government had determined upon sending an expedition for the chastisement of the Butan savages in south-eastern Formosa.[1] A small vessel belonging to some merchants in Japan was some time previously wrecked on the coast of that island, and the crew, with only a few exceptions, was massacred by the natives. Failing to obtain any satisfaction, it was decided to punish the murderers, and to sack and burn their towns and villages.[1]

The question, however, of getting their troops to Formosa was one which presented the greatest difficulties to the Japanese War Department, but Captain Brown, who had navigated the tender round the whole of the unsurveyed coasts of the country without accident until April 1874, stepped into the breach. Without vessels which would serve as transports and without any subject of the Mikado capable of navigating a steamer so far as Formosa, the Imperial Government seemed in a helpless position.

A large vessel, named the *Kuroda*, which had been

95

purchased by some native merchants was discovered, and requisitioned, against the will of the owners, by the War Department, and Captain Brown volunteered to take charge of her.

But diplomatic difficulties supervened. Formosa was supposed to belong to China; and Great Britain being on terms of peace with that empire, Sir Harry Parkes objected to a British subject's taking so prominent a position in a hostile expedition against a friendly nation. Captain Brown was duly warned by his minister and threatened with the penalties declared by law against those acting against a foreign state, in contumacy to the Queen's representative. Nevertheless he still persisted in his resolve to assist the Japanese. Taking up his quarters on board the *Kuroda*, he was twice ejected from her by her owners, but the Japanese authorities eventually succeeding, he sailed in her, taking seven or eight hundred troops, for Formosa.

That he thought his forcible detention by the British authorities probable, is shown by the following letter received by me:

Saturday morning.
11th April 1874

My dear Mr Brunton,

Can you tell me whether Captain Brown, of the Lighthouse tender, has left in the *Kuroda*? He told me that he was going to Amoy and was to take General Legendre there, but I have certain information that that is not the destination of the *Kuroda*.

Yours very truly,
HARRY S. PARKES

Having with the greatest vigour and diligence carried troops backwards and forwards between the south of Japan and Formosa, and the expedition having accomplished its object in chastising the Formosa natives, Captain Brown became the object of the most enthusiastic gratitude on the part of the officials. Presents of all kinds were showered upon him, a place with a high salary was made for him in the Government service, and special honours were bestowed upon him. At the present day he acts as Honorary Japanese Consul at Glasgow, superintending the completion of the numerous vessels constructed in that port for Japan.

CHAPTER TWENTY-THREE

VICISSITUDES

THE CONSCIENTIOUS and efficient conduct of new work in Japan was a task which presented the most perplexing difficulties both to employers and employed. Their high pay, their different mode of living, their want of disciplinary power, and the knowledge of the Japanese that foreigners were more or less indispensable to them, rendered their European assistants most impracticable and difficult to deal with. Resignation, insubordination, absence from duty, drunkenness and other aberrations of conduct among Europeans employed in the Japanese Government service, became frequent and distressing. On the other hand, the semi-ignorance of the native servants of the Emperor, and the self-esteem, untrustworthiness, craftiness and corruption of the Japanese underlings rendered cooperation by an honourable foreigner with them extremely irritating.

In dealing with the Japanese officers who considered themselves my employers, it soon became apparent that I must perforce choose one of two lines of conduct.

The first choice, which promised quietude and repose, was to let things take their course, give advice when asked for, feeling undisturbed if this was neglected and to become imbued with the Oriental estimation of the valuelessness of time, allowing nothing that hinders progress to perturb or annoy. Such was the method, in the Seventies, by which the European could become a favourite with his Japanese employer; and such, it is to be feared, was the mode adopted by most of those entering the Government service, with the natural result that much of the early work done caused scandal and opprobrium.

The other method was an insistence by the European official employed, of a due enforcement of his directions; he being, from the position in which he had been placed, responsible for results. While such a line of conduct would go far to ensure the execution of the appointed work, according to the ideas of the person in charge, it was almost certain to create friction, ending possibly in rupture of relations. It was evident that the Japanese had made up their minds to make what use they could of their foreign servants, but in no case to have them become masters, or to invest them with any power. They would hold them in the position of advisers or instructors only, without the authority to direct.

Besides feeling the indignity of remaining in such a position, I formed a clear opinion that it would certainly tend to an unsatisfactory execution of the work. I came therefore to a decision, that, at whatever personal discomfort and self-sacrifice, I should assert my position as the responsible conductor of operations.

A further consideration presented itself to my mind, viz: that while other works undertaken by the Japanese Government may have been their own domestic concerns, the construction of lighthouses, which had been intrusted to me, was a work covenanted for in solemn treaties, and was in the interests of humanity at large.

I felt myself therefore responsible to foreign Powers as well as to my immediate employers.

I, therefore, had no hesitation in demonstrating to those from Tokio, appointed to cooperate with me, that after work had received official sanction, it was my intention to take the direction of affairs actively and personally.

Fortunate in having the full concurrence of Sir Harry Parkes in the adoption of this attitude, I was assisted by him with his advice and with his great influence with the Mikado's Government.

On the other hand, I was fully desirous that the reasons for my actions should be fully understood by those appointed to work with me. I spared neither time nor pains in giving them the fullest explanation.

I am obliged to confess however, that I met with only partial success. Independent official instructions were given without reference to my schemes and sometimes these actually clashed with my directions. Information was withheld from me with which properly to conduct the work, and with which I should have been furnished.

This dishonourable secrecy was especially felt in regard to the financial side of lighthouse construction. The cost of the work was not allowed to be known. I began with the system, usual in Great Britain, of permitting no payments to be made without a certificate from me, as Engineer-in-Chief, that such were really due. I proposed also keeping a detailed account in England of the cost of each piece of work which I had in hand. When I informed my official superiors of my plan, they, at first, seemed to acquiesce in it, and sanctioned my engaging an accountant to keep the books. But in about eighteen months, payments began to be made by the officials without my knowledge, and, eventually, all operations were proceeded with without my obtaining any detailed information of the amounts expended on them.

To the wholesale system of corruption which permeated the public service in Japan at this era, may be traced the desire for what I believe to have been guilty reticence. Had the engineer's certificates remained the sole authority for payment, manipulation of charges or peculation of funds would have been wholly prevented. But this did not suit the desires of those persons to whom public service meant private gain, and so payments were made independently of my authorisation.

The conduct of the work of lighthouse extension was, further, rendered most difficult and wearisome, by the constant changes which were made in the personnel of the native staff. Had I been permitted to cooperate with one or two select and intelligent Japanese, I no doubt could have in time imbued them with good practical ideas of the methods in which the work should have been carried out.

But during the time of my service in Japan, there were no fewer than fifteen heads appointed over the lighthouse office, each one with his own preconceived notions, and none with any previous experience or knowledge whatever of the undertaking he was sent to direct. On the 20th of December, 1869, I was introduced to Ito, now the Marquis Ito, and four times Prime Minister of Japan, who was then an under-officer in the Finance Department, and who, it had been arranged, would take a general supervision of my work. Having under him Saito, a gentleman of good intelligence, who had been in contact with my work to some extent from my arrival, things moved along smoothly and satisfactorily during this time.

Though anxious that this arrangement should continue, I was annoyed to learn that another new office-holder had been appointed to take charge, Ito introducing him on the 24th of March, 1870. A few weeks later, I was informed two young officers would act as joint chiefs of the Lighthouse Department.

Ignorant, obstinate, haughty and imperious in their

manner, this pair gave me while they held their positions no end of trouble. Not only did they direct affairs after their own way in every branch of the service, but they deputed their authority to a number of minor underlings, who acted independently of the chief engineer's, and even took upon themselves to give him instructions as to his procedure.

At this time, Okuma, already mentioned as one of the trusted councillors of the Mikado, was Minister of Finance. Understanding that the lighthouse service was under this Department, I appealed to him regarding the treatment I was receiving at the hands of those set over me. I pointed out their inexperience and their incompetency and the danger to the work which their exemption from all control involved. I asked what use it was to have a European staff, if opportunities for work were to be thwarted by persons who were necessarily without the required knowledge. I complained further of the indignity of having to receive orders from men in subordinate positions. To these representations, Okuma replied that a Department of Public Works would shortly be organised, having Ito as its head, his subordinates being chosen, so far as possible, from Japanese who had been educated abroad.

In February 1871, The Department of Public Works was established; but for some reason, the lighthouse service was not placed under it. On the 4th of May, I was informed that the matter of coast illumination was partially under the Home Office, and partially under the Public Works Department, the official work being transacted by the former, while the latter assisted with advice when asked for. The reason for this, probably, was that at this time the newly formed Public Works Department had its hands fully engaged with the Railway then begun between Yokohama and Tokio, and it was considered unsafe to overburden it too early in its career. It involved, however, the rentention of the twin directors in the Lighthouse

Department until September, when it was entirely transferred to the Public Works Department and these two gentlemen left.

Besides being at the head of this Department, Ito was appointed a Councillor of State, and the active direction of Public Works devolved to a great extent upon another officer. Although educated in England, and speaking English fluently, this latter person was a Japanese of the old type, stubborn, self-opinionated and with the most lofty conception of the high standard of intelligence of his race. His instructions, given with a peremptoriness and ostentatious disregard for my views, rendered intercourse with him extremely irksome. It was this gentleman who, during the absence of Ito and myself, arranged for the building of the additional steamer for the lighthouse service, which as has been already mentioned, was not required, and which involved the expenditure of a large sum for no conceivable purpose.

Happily for me, a gentleman named Sano[1] was appointed as the 'Chief' of the Lighthouse Department, who remained until the end of my first period of service, that is, until I returned to England on absence of leave. Always suave and politic in manner, and lofty ideals and generous sentiments, it is impossible to recall the months of intercourse with this true samurai without a sense of pleasure. Acting, probably, on instructions, he was, it is true, constantly asserting his position and controverting my desires, but this was done by quiet and dignified assertion or argument, and it was only on very few occasions that he succeeded in gaining his way.

On one occasion, it having been arranged that some additional lighthouses were to be erected in the Inland Sea, I proposed to Sano that he should take advantage of the opportunity afforded by the next trip of the lighthouse tender to visit and examine the sites. For some reason he objected to this, and informed me that he had instructions from Tokio that I was not to do so. Conceiving this to be an instance of arbitrary obstruction, for which no reason could be given, I appealed to Sir Harry Parkes. That gentleman at once expressed the opinion that the proposed visit was a mere professional detail with which the officers had no right to interfere, and I should disregard their prohibition of it, and, if any blame was laid upon me in consequence, I was to refer the complainants to him. Needless to say the visit was made and no censure was expressed.

On another occasion, when many lighthouses had already been established at points along four thousand miles of coast-line, I desired that definite arrangements should be made for regularly replenishing the stores of materials required for the maintenance of the lights, such as oil, wicks, glasses and cleaning stuffs. Following the system adopted in Scotland, I had provided tanks and storerooms at each station sufficient to maintain one year's supply. In Japan, owing to the absence of internal communication, to the great length of coast to be traversed and the difficulty of replacing the tender, should any accident happen to her, I came to the conclusion that not less than one year's supply of stores should always be on hand at each lighthouse.

Sano, however, in the exercise of his judgement, considered that a six months' supply was sufficient. This was a vital point in my opinion, and as the British Minister had at this time left for home on leave, I had to contest it unaided. The only reason given for differing from the foreign engineer was that having a large quantity of stores on hand, the lightkeeper might be tempted to waste them. Yet to me, since returns of the expenditure of stores had to be sent to the office monthly, this contention appeared childish.

Being resolved that my decision should be accepted, I had given instruction to my European assistants to ship one year's supply, and before the Japanese were aware of what had been done, a large portion had been transferred to the steamer. On Sano discovering it, he ordered that one half of that which had been shipped was to be returned. On my pointing out the practical impossibility of doing so, Sano became indignant, as he said, at being forced to acquiesce in what he had decided against; but made no further objection. Thus the lighthouses received their one year's supply.

At another time Sano, with his old-fashioned notions of etiquette, proposed that on board the lighthouse tender the Japanese at meals should sit on one side of the table and the Europeans on the other, so that those of each country could be placed at table in order of their rank. But on my objecting, this whimsical idea was heard no more of.

When it became necessary to send money to Messrs Stevenson to enable them to obtain new light apparatus required, Sano evidently considered he was doing a very smart piece of business, when he asked me to become security for them. Naturally refusing, I explained the position those gentlemen held in the estimation of their countrymen. Then Sano asked me if I would give a letter guaranteeing that Messrs Stevenson held this position. To this I agreed, and on receipt of the letter, the money was sent.

Such were the foibles of the man, but at bottom he was a man of good heart and good understanding.[1] Some years after his connection with the Lighthouse Department had been severed, he invited me, together with other servants of the Japanese Government, to a banquet at one of the Government Offices in Tokio. Before sitting down, he called me on one side and made profuse apologies for the way in which, in his ignorance, had thwarted my desires and obstructed my progress. He acknowledged that he had been brought to see how greatly he had been in error. It was this gentleman who, at the end of the Satsuma rebellion[2] of 1877, formed the 'Philanthropic Society', which established some hospitals, and into these, for the first time in Japan's history,[2] and much to the astonishment of the native population, the wounded and sick of the enemy's troops were admitted for treatment. This Philanthropic Society was ultimately merged into the Red Cross organisation. Altogether Sano's character was one which in the midst of civilisation would be regarded as fine and noble.

Looking at the matter generally, the treatment of their European employees by the Japanese Government was marked by a spirit of unenlightened distrust and jealousy. This while preventing the free exercise of their abilities, rendered their period of service either one of enforced activity, or continued struggle.

Instead of having free scope for their talents, the foreign servants of the Emperor were swathed round so tightly with the cramping bonds of suspicion and jealousy that they were unusually helpless. In most cases they had little communication with the heads of the Government. Their actions were controlled and watched by subordinates. The carrying out of the smallest detail in their work was criticised and interfered with. All suggestions had to be laid before men in inferior position, and with them, the decision rested as to whether they were worthy of being referred to the head of the Department or not.

Against this grave inquisition I fought strenuously, and with the aid of Sir Harry Parkes I succeeded in retaining the active control of the works so far that when completed they were generally regarded as successful. On the contrary many Europeans less fortunate saw the undertakings they had been engaged to direct mismanaged and bungled, and their own reputation in consequence seriously damaged.[3]

NECESSITY, THE MOTHER OF INVENTION

HAVING at the end of 1869 made due arrangements for the erection of lighthouse towers, etc., on the sites chosen for them, and the work proceeding with fair rapidity, I was distressed to learn that the vessel bringing out the illuminating apparatus to Japan, had some months previously been totally wrecked on the coast of Formosa. The Japanese at once gave instructions through Sir Harry Parkes to re-order from Europe all that had been lost.

That strenuous personage, the British Minister, could not brook the delay which this unfortunate accident would inevitably cause. With the utmost vehemence, he urged me to exercise my ingenuity in order that temporary lights might be shown on the more important headlands. Thereupon I attempted the not very easy task of 'making bricks without straw'. But, as usual, necessity proved herself a good mother.

Having discovered some skilful Japanese coppersmiths, I placed the design of a lamp which I had made before them. In a few weeks time I was pleased to find that these artisans had succeeded very creditably in making several after my model. Obtaining some masthead side light lenses from the ship chandlers' shops in Yokohama and Hong Kong, I placed these lamps, burning kerosene, in their focus. By arranging them in groups of six or eight, in improvised lanterns, a light of sufficient power was attained to penetrate considerable distances. From San Francisco, I imported a number of the large headlights, or parabolic

reflector lamps, used in the front of American locomotives. In the same way, by grouping these in lanterns, very powerful illuminants were produced.

Next in order was the erection of temporary towers
of wood. I placed on these the extemporaneous apparatus obtained and was enabled, in a very few months,
to announce the establishment of six or eight sure-
burning lights on these headlands, where, for the safety
of foreign vessels, illumination was most required.

A characteristic occurrence marked the issuing of
the announcement of these safeguards to navigation
along the coast of Japan. Having drawn up the notification after the system adopted in Great Britain, in
the form of 'Notices to Mariners', giving the latitude
and longitude of the headlands, their height above sea,
the distance at which the lights were visible, and information regarding dangers in the vicinity, I naturally
assumed that the engineer, as being responsible for
the document, should sign it. To this the native officers
objected. They were willing, they said, to accept the
assistance of a European, but they were not prepared
to publish to the world the fact that any Department
of their Government was under the control of a
foreigner.

Feeling that this was a matter of some importance,
so little being known of the Japanese at this period,
that announcements coming from them alone would
probably not be received with the confidence necessary
to make them useful, I referred the matter to the British
Minister. He at once gave his judgement that the
'Notices' should bear the name of the engineer by
whose direction and under whose responsibility the
lights were established.

Even to this judgement the Japanese showed an extreme reluctance to accede. In the end a compromise
was made, by which the chief servant of the Emperor
in the Lighthouse Department signed the announcement as 'Commissioner of Lighthouses' and I as

'Engineer'. Foreign navigators thus had ground for greater confidence in the trustworthiness of the announcement.

CHAPTER TWENTY-FIVE

BUILDING SHIPS

TWO OR THREE Lightships being urgently required, I, with a boldness which, looking back, I now regard as audacious, set to work, having them built under my own guidance. Luckily dropping upon an intelligent ship's carpenter in Yokohama, who had been brought up in a shipbuilding yard in England, I engaged him, placed the drawings in his hand, and set him to work with Japanese workmen and native material. So promptly was his work executed, that on the 24th of June, 1869, Japan's first lightship was successfully launched. Having been fitted out precisely after the methods of vessels of the same character in England, she was moored in her position on the 28th of September of the same year. The lightships were seventy feet long, and of 130 tons burthen, and fitted with two decks and two masts. The light at the mainmast was sixteen inches in diameter and forty feet high. The native woods used were *keyaki* and *hinoki*.

An incident occurred at this ship launch which displays another feature of Japanese character.

For convenience sake, the vessel had been built at a place on the shore, off which there was not much more depth of water than was sufficient to float her. When she slid down the ways, her stern post was driven into the soft bottom and she was stopped before being entirely free. Her position was that her stern was down, stuck in the mud, and her bow was up on the ways. When the tide fell, she might heel over or otherwise damage herself.

I took the resolve at once to have the ways removed from below the hull, so that either the ship might float, or her whole frame might rest solidly on the flat mud. But now a new obstacle presented itself. Unknown to me, the Japanese in charge had promised the native ship's carpenters that after the vessel had been launched, they were to be treated to *sake*, which, though a beer, made by brewing rice, is intoxicating.

Away rushed the workmen for their promised drink, the officers, heedless of what had happened, giving it to them in such quantities that numbers of the men were perfectly incapacitated. Turning the useless ones away by actual physical force, and forcing the others to return to their work, I eventually succeeded in floating the vessel. This occurrence gives a good instance of the want of ordinary capacity to appreciate palpable positions of difficulty and the proneness to act independently, against which Europeans had in those days to fight.[1]

Floundering along in this way, spending valuable time in vain reasoning with impractical politicians and office-holders, and meeting with continual obstruction from them, the progress made under the circumstances seems wonderful.

CHAPTER TWENTY-SIX

AUDIENCE OF THE EMPEROR

WHEN after about four years service, I applied for leave to return home, there were fourteen lights shown on the coast of Japan, two lightships and twelve buoys moored off shoals, and three beacons erected on sunken rocks.

So far did the Government appreciate this work, that during the same year, 1872, the Dai Jo Kuan,[1] or great council of the Government, issued the following:

NOTIFICATION

'Lighthouses are very useful for navigation. Formerly Japanese were in the habit of burning wood fires, at the expense of the surrounding people, which was very expensive, and these fires were often put out by the wind and rain and so they became a great danger to navigation. Therefore there will now be constructed lighthouses where they are found necessary and useful. All the original wood fires must, from this time, be discontinued and put an end to.

Any local Government or authority therefore, in the country, desirous of having a lighthouse on any point must make application to this office.

Dai Jo Kuan

15th day 5th month.
4th year Meiji.'

Some months previously, together with other heads of Departments in the employ of the Imperial Government, I was honoured by being formally presented to the Mikado.[1]

The first Europeans[2] his Majesty ever saw are believed to be the Duke of Edinburgh and his suite, who, visiting Japan in the *Galatea* in 1869 was received in public and private audience by him. Sir Harry Parkes was presented in May 1871, on the occasion of his departure for England on leave, and he also had a private audience.[2] He seized this opportunity to read the Mikado a lecture on the necessity of placing full confidence in the foreigners in his employ. He desired the Emperor not to be satisfied with the superficial knowledge of arts and sciences which his subjects might attain by hasty travel abroad.

During the same year General Capron and a large staff of assistants arrived from the United States[3] for the purpose of colonising and developing the Northern island of Yezo, then peopled by a distinct race named Ainu.[4] On his arrival he was presented to the Mikado, when, His Majesty having shortly addressed him, he made a lengthy reply. This being translated into Japanese was read to the Emperor. On being published in the Court Journal, it was discovered that a word had been put into General Capron's mouth which he certainly never uttered. He was reported as having described himself as the *bishin* or 'insignificant servant' of His Majesty.[5] The substitution of this word, for the equivalent of the simple personal pronoun 'I', being a liberty taken by the Court officials, was made presumably, in order to show their superiority over him and was much resented by him.

When I was presented, the Mikado's speech and the reply which I was expected to make were written out in Japanese. An officer having a fair command of English gave me the gist of their contents, making the request that I should put both into a proper form in my own language. The wording, however, was a matter of great punctiliousness on the part of this gentleman, it being his desire that the Emperor of Japan should not commit himself to anything definite, but, at the same time, should convey his wish to add to works of improvement. The indiscretion committed in the case of General Capron, was in this way, avoided, but that I should have been requested to address the Emperor and also to dictate the words of a reply for which I was in no way responsible, was one of those curious absurdities which seem to be the natural product of a culture and etiquette based on slavish adherence to Chinese precedents.

On the 17th of November at 8.30 a.m. I was driven in a carriage sent for the purpose to the Mikado's palace. Entering one of the many pavilions of which the old structure was composed, I was kept there till 10.30 a.m. sipping tea and smoking after the fashion of the country, while conversing with the officers present.

When the time arrived for the ceremony, Ito, then Minister of Public Works, introduced me. The Mikado sat on a raised dais covered with red silk and wore a magnificent embroidered robe of the same material. Neither myself nor Ito entered the compartment in which His Majesty was, but stood in a recess outside, about six or eight feet distant from him. The Emperor bowed in a stately manner when I came in view, but his countenance was perfectly rigid and stolid, not the slightest appearance of a smile of welcome breaking through its immobility.

Having read his address in Japanese,[3] and Ito having read a translation of this in English, the Emperor listened to my reply, which Ito again translated into Japanese.

The Emperor said:
'The work of lighthouse construction on the coast has been successfully proceeded with by your assistance.

'And by the aid of this work, the dangers of navigating the coast are to a great extent lessened.

'I fully appreciate your efforts in the work of constructing these lighthouses already completed, and the diligence with which you have carried out, so far, the work entrusted to you shows your merit to be of no shallow kind.

'I trust that by continuing to act in the same line of conduct, I may increase the facilities which these works afford.'

To this I replied:

'I beg to express the greatest gratification at the flattering appreciation of my services which your Majesty has been pleased, on this occasion, to give expression to.

'In an insular country such as Japan, on the coasts of which there are many dangers and difficulties for the navigator, the construction of lighthouses, and the establishment of a well-regulated lighthouse system is a work of the utmost importance and advantage.

'My efforts, while in your Majesty's service will always be directed toward that end, and I trust that in future with the assistance of your Majesty's officers they may meet with success.'

The ceremony of audience being over, I bowed myself out of the august presence.

A banquet, attended by many of the higher officials of the Government was given in the evening in the Mikado's palace, to which I was invited;[6] and many felicitous speeches were made, in which as is usual in such circumstances, all differences were sunk and congratulations exchanged.

CHAPTER TWENTY-SEVEN

THE GREAT EMBASSY TO THE TREATY POWERS

ON THE 23rd of December, 1871, the most important embassy[1] that had ever left the shores of the Empire of Japan embarked on board the Pacific Mail Steamship *America*, primarily for the United States but with the intention of visiting every nation with which Japan had treaties.[1]

The purpose for which the voyage was undertaken was given out to be the acquisition of a more intimate knowledge by Japanese statesmen of foreign civilisation....

Before its departure, the Mikado gave an address, using the following language:

'After careful study and observation, I am impressed with the fact that those are the most powerful nations whose peoples have most diligently cultivated their minds and developed their countries. Thus convinced, it is my duty as a sovereign to lead my people wisely and to assist them to diligently and unitedly attain such beneficial results.

'If we would profit by the useful arts, sciences and conditions of society prevailing among more enlightened nations, we must either study these at home or send abroad an expedition of practical observers to foreign lands to acquire such information as our people lack.

'Travel in foreign countries will increase your store of knowledge, and although some of you may be advanced in years and unfitted for the vigorous study of new ways, all may bring back much valuable information.

'We lack institutions of female culture. Our women should not be ignorant of those principles on which the happiness of life depends. On the education of mothers depends the early education, in their progeny, of those intellectual tastes which are of such vast importance. Liberty is therefore granted wives and sisters to accompany relatives on foreign tours, in order that they may learn the best manner of female education and introduce it in the country.[22]

'With diligent and united efforts among all classes, we hope to attain the highest degree of civilisation, and shall so obtain power, independence and respect among the nations.'

The Japanese envoys, in so announcing the object of their mission, obscured its real purpose. As the date on which it was permissible to make a revision of the foreign treaties was the 1st of July, 1872, the Government in Tokio made up its mind that it would listen to no suggestion of revision, unless the extra-territorial clauses, which precluded Japanese jurisdiction over resident aliens in the country, were annulled. It was therefore mainly with the object of influencing foreign Governments at their sources, towards this end, that the mission was undertaken.

The Embassy gave an exceedingly complete representation of New Japan. Besides Iwakura, one of the nobles in the Mikado's court in Kioto and one of the most notable figures in the restoration of 1868, Kido, Okubo, Ito and Yamaguchi[3] - all heads of Departments and men of mark, were about forty-five secretaries, commissioners and attachés.

Landing in San Francisco on the 15th of January, 1872, the Japanese were warmly welcomed by the Mayor and local authorities. Banquets and receptions, with speeches of exaggerated compliments, were daily recorded, and the progress made by Japan and the promise of future progress largely dilated on.

It is notable that no mention was made of the real

115

object of the Embassy, the revision of the treaties; but Ito, in a notable speech, gave utterance to some remarks which are worthy of quotation. He said:

'Our mission under special instruction from his Majesty the Emperor, while seeking to protect the rights and interests of our respective nations, will seek to unite them more closely in the future, convinced that we shall like each other more when we know each other better.

'Although our improvement has been rapid in material civilisation, the mental improvement of our people has been far greater. Our wisest men, after careful consideration, agree in this opinion. By educating our women we hope to secure greater intelligence in future generations. With this end in view, our maidens have already commenced to come to you for education.

'Japan cannot claim originality as yet, but it will aim to exercise practical wisdom by adopting the advantages and avoiding the errors taught her by the history of enlightened nations.

'The red disc in the centre of our national flag shall no longer appear like a wafer over a sealed empire, but be henceforth in fact what it is designed to be, the noble emblem of the rising sun moving onward and upward amid the enlightened nations of the world.'[4]

Proceeding through the Pacific coast states by rail, the mission found itself snow-bound at Salt Lake City. It has been reported that the Mormon chief, Brigham Young, sent a message to the principal ambassador requesting him to call. Pointing out that the etiquette of his country forbade his making calls until he had received a visit, the ambassador asked why the Mormon chief could not call on him. He was told that he was at that moment confined to his room, in charge of an officer of the United States Government.

Having heard of the efforts being made by the Federal Government to put down bigamy in Utah, the Japanese Ambassador at once recognised the position, and is reported to have said:

'We came to the United States to see the President and we cannot insult him by calling on a man who has broken the laws of his country and is under arrest.'

Arriving in Chicago in February 1872, the Embassy were witnesses of the great conflagration which had shortly before nearly annihilated that city. Subscribing among themselves $5,000 gold towards the Mayor's fund for the alleviation of distress occasioned by the calamity, they were gracefully thanked for their 'personal sympathy'. They then proceeded to Washington, which city they reached on the 29th of February.

The Congress of the United States had voted $50,000 gold for their entertainment, and to General Myer was entrusted their comfort and safety. They entered on a round of banqueting and complimentary speech-making of an altogether phenomenal character.

Having an audience with the President, General Grant, on 4 March, their letter of credence from the Mikado was read to him. In this occurred the following:

'The period for revising the treaties now existing between ourselves and the United States is less than one year distant. We expect and intend to reform the same, so as to stand on a similar footing with the most enlightened nations, and to attain the full development of public right and interest.'

In this passage we obtain the first indication of the real object of the Embassy. The 'desire to stand upon the same footing as the most enlightened nations',[5] when the treaties should be revised, was the real and true reason for its despatch. The Congress of the United States gave them a formal reception, when more complimentary speeches were made, and amid much ceremony, each member of the Embassy was introduced to the chief members of the Cabinet before a house crowded with the beauty and fashion of the Capital.

CHAPTER TWENTY-EIGHT

HOME AGAIN WITH THE JAPANESE IN ENGLAND

THE JAPANESE Embassy after feasts and fêtes in most of the large towns in the Atlantic Coast States, sailed from New York for England early in August 1872, arriving in London on the eighteenth of that month.

Having obtained leave of absence, I left[1] Japan on the 24th of April of the same year, and arrived in London on the 18th of June. I lost no time in calling on the Embassy, which was located at the Buckingham Palace Hotel, being received by its various members with the greatest cordiality.

Sir Harry Parkes was constantly in attendance on them, while General Alexander was specially appointed as the representative of the Queen to see to their comfort and well-being.

They were received by the Prime Minister and other members of the Government and were accorded an audience with her Majesty; but the reception of the Japanese in England was not marked by any of that effusiveness of public oratory which distinguished their visit to the United States. Absolutely nothing was made public of their intercourse with the officers of the British Government, which under any circumstances could not have been of much importance; but their desires seemed to be almost wholly pointed to an examination and study of the industries of the country.

While Iwakura, Kido and Okubo were escorted by General Alexander and Sir Harry Parkes to the great Government establishments at Woolwich, Chatham, Portsmouth, etc., Ito, as Minister of Public Works,

118

together with a number of intelligent young attachés, put himself under my wing, and by me was introduced to a large number of manufacturers. Twenty-eight places of business in London were visited during the month of September. In these were comprised the most diverse occupations, such as the making of candles, skivers, glue, gelatine, bricks, plumbago, matches, India rubber, Portland cement, watches and clocks, pottery and terra-cotta, gunpowder, paints, colours and varnish, besides leather-dressing, tanning, dyeing and bleaching, meat-preserving, etc., etc.

The various processes seen were minutely studied and carefully noted by the chief and his secretaries. The owners of the factories in every case, not only assisted by giving every explanation in their power, but in many cases hospitably provided lunch for the visitors.

Arriving at Birmingham at the end of September, Ito with his attachés and myself visited the most important works in that city, such as the small arms factory, Chance's glass and chemical works, Elkington's Electroplating, Gillot's steel pens, etc. In Manchester and Liverpool we saw Tate's sugar refinery, Houldsworth's cotton mills, Hayle's calico printing works, Whitworth's engineering establishment and Sharp Stewart's locomotive works. At Edinburgh we joined the heads of the Embassy and Sir Harry Parkes. Here the Commissioners of Northern Lights invited the visitors to visit the Bell Rock in the lighthouse tender *Pharos*. This they did on the 16th of October, though under rather depressing circumstances as regards weather. It rained the whole day and blew so hard that it was impossible to land at the lighthouse. After dinner on board, during the return journey, felicitous speeches were made concerning the construction of lighthouses in Japan, and I was duly complimented on what had already been accomplished.

The Embassy visited some of the large shipbuilding

119

yards on the Clyde, and then came on to Newcastle,[2] where they had an opportunity of inspecting the works of Sir William Armstrong, from which they have since obtained so many vessels and munitions of war. The hydraulic machinery and appliances with which his whole establishment was fitted were explained to them by Sir William Armstrong himself. One of the first Gatling guns ever made, which had ten barrels and fired 250 shots per minute, was shown at work.

From Mr Bell, of Rushpool Hall, Yorkshire, I received an invitation for myself and Ito to visit the iron works in the Middlesbrough district, owned by the well-known firm of Messrs Bolkow and Vaughan.

The question of iron smelting and rolling was one which, in view of the large quantities of ore which were known to exist in Japan, was of exceeding interest to Ito. Mr Bell, while acting as host, most hospitably afforded Ito every item of information most fully, and promised his assistance should the Government think of developing the mines of Japan.

The Embassy at Sheffield had an opportunity at Cammell's Steel Works of seeing the Bessemer process of making steel, as also the rolling of armour plates, and were most lavishly entertained by the Messrs Wilson, the managing partners of the firm.

The late Duke of Devonshire invited the Embassy to pay him a visit to Chatsworth; and all its principal members, accompanied by Sir Harry Parkes and myself, proceeded there on the 30th of October.

We were received by the Duke, Lord Charles Cavendish, Captain Egerton and a number of ladies of the family. The Duke personally conducted the party through the whole house, and to the kitchen and wine cellars, where a glass of beer brewed by his grandfather was offered to each of the visitors. He also led us to his stables and gardens, describing the value and interest attached to every article of note which was passed. Lunching in the chief banqueting hall, off a golden service of plate, the Embassy declared that no day had so much impressed them with admiration of English institutions as that.

After visiting various cutlery establishments and attending the Cutler's feast in Sheffield, Ito and I again detached ourselves from the Embassy in order to enjoy an invitation given to us by Mr James T. Chance of Four Oaks Park, Birmingham. Accepting the hospitality of this well-known gentleman for three days, Ito received from him much valuable information regarding the manufacture of glass, which it was much desired to introduce into Japan, and was also taken to celebrated buildings in the locality and to a meet of the foxhounds.

Ito then rejoined the Embassy, which at this time had arrived in Birmingham, and on the 4th of November I left him and returned to London.

Shortly after my arrival in London on this occasion, much to my amusement and surprise, I received a visit from a Japanese merchant who had been sent to England in order to purchase iron pipes for the Yokohama water supply! I was naïvely asked by him what size he should order them. The absurdity of sending a man from Japan to England to purchase cast iron pipes, when there were agents on the spot who could have procured them, and the quaint innocence of his leaving without knowing the size they were required to be, is quite illustrative of the vagaries of the Japanese mind of a generation ago. Having none of my figures beside me, I could give him no satisfactory information. Whether the pipes were purchased or not, I never heard.

Receiving many commissions from different members of the Embassy either on account of the Imperial Government or for their own private purposes, I was kept busy until they left, on the 14th of December, for Paris, having been four months in the country.

Advantage was taken of the presence of Ito, the

Minister of Public Works, to arrange for closing the accounts of the money sent from time to time, to the Board of Trade for lighthouse apparatus. The author having gone through these at the Board of Trade and certified to their correctness, it was arranged by Ito that the balance in their hands amounting to about £2,000, should be transferred to Messrs Stevenson for the purchase of such new apparatus as might be ordered from them.

It is a somewhat curious circumstance that this arrangement, the most obviously convenient for all parties, did not meet with the approval of the Finance Department in Tokio. They desired the balance to be returned to them, as it was part of the moneys paid by the late Taikun's Government, and they would remit to Messrs Stevenson fresh money as required. But so far as I am aware, and I feel sure that I should have been made aware of the circumstance had it occurred, no expression of thanks has been made to the Board of Trade for the care and unstinted trouble which, to my personal knowledge, the officials of that Department gave the matter of the Japan Lights.

CHAPTER TWENTY-NINE

JAPANESE PETROLEUM

MANY AND DIVERSE were the projects which were discussed by the Ambassadors during their stay in London. The development of the mineral resources of the country was gravely spoken of, and information sought regarding the establishment of smelting and rolling mills for iron and also for the refinement of petroleum. No definite steps however, were taken with either project. In regard to the latter of these, I felt a particular anxiety that some decision should be come to, as the supply of vegetable oils for consuming in the lighthouses was found to be most uncertain in Japan, and mineral oil had become, by recent appliances, an excellent substitute. (It was just at this time that burners invented by Captain Doty for consuming paraffin oil, had been introduced into the Scotch lighthouse system, and gave remarkably advantageous results, the amount of light being about doubled, at one half the cost of the colza oil hitherto used.) Having obtained Ito's authority to obtain these burners for the Japanese lighthouses, and knowing that an excellent crude petroleum could be had from wells in various parts of the country, I lost no opportunity of impressing upon the members of the Embassy the advantage to be derived from erecting a refinery at least of such capacity as would supply a quantity of refined oil sufficient for their use.

But as the following letter shows, delay was decided on:

'Office of the Japanese Embassy
LONDON
14th December 1872

Dear Mr Brunton,

Referring to the purchase of machinery for the distillation of petroleum, I think it would be advisable and prudent to wait until you return to Japan, and until the exact place for establishing this is decided on. If the purchase were made now, the machinery might not be adapted to the circumstances. For this reason, I would desire you not to give any order now, but so to arrange matters that you can order it at any future time after the position is settled on.

As to the employment of oil refiners, these need not be actually engaged until the machinery is purchased, but you can make a selection of the men now with the assistance of your friends.

I remain,
Yours faithfully,
Hirobumi Ito.'

Notwithstanding the avowed intention, as expressed, of utilising the petroleum springs, and the fact that immediately on my return I urged the matter on the Imperial Government, continuing to do so on many occasions for several months, the matter never received consideration, and no decision was come to. The only alternative was to obtain oil from abroad. Indeed, supplies were secured from Scotland up to the time of my leaving the country.

In April 1874, when in the vicinity of one of the lighthouses, during one of my journeys in Japan, I visited some petroleum wells, situated in the province of Totomi,[1] about 100 miles from Yokohama. The wells are forty-seven in number. From only seventeen of these was oil at that time (April 1874) obtained, and, from them it was got in small quantities only a few miles from the sea coast.

The wells were in the form of square holes, four feet on each side, and lined from top to bottom with wooden sheeting. The greatest depth to which any of them were sunk was five hundred feet, the geological formation through which they passed being first surface soil, second clay, third soft sandstone, fourth hardish blue clay-stone, underneath which is loose sand or gravel in which the oil is. The wells were dug by men with shovels and buckets down to the bottom. The vapours from the oil were exceedingly noxious, and air was pumped down to the men through a wooden trough, by means of exceedingly primitive and roughly made bellows worked by three men with their feet. There had been several instances of failure on the part of these bellows, with the result of the men below being suffocated. There were no means of communication between the man below and his comrades on the surface, and as he was expected to remain down five or six hours at a stretch, his position was not at all an enviable one. The oil was raised in buckets by four men, hand over hand.

The only attempt to refine the crude oil was by a single distillation, in which a portion of the tarry refuse was removed. At its best the product of the wells was of little value as an illuminating agent. It was said that about 600,000 gallons per annum of crude petroleum were raised by these primitive means. Some of it was sold locally, but most of it went to Tokio.[1]

At Ito's request, I selected and engaged a number of artisans principally for ship repairs, and made arrangements for them to proceed to Japan.

CHAPTER THIRTY

WOMEN'S EDUCATION IN JAPAN

IT WAS ALSO the desire of the Embassy to engage a few English ladies to go to Japan, to teach Japanese women household duties as known and practised in Great Britain. I was therefore deputed to engage for a period of three years, three ladies, one a matron at $150 per month, another a governess at $100 per month, and a third a seamstress at $60 per month. These women were to have free quarters in Tokio and their passages out and back paid.

To fully understand this matter, it will be necessary to have some knowledge of the position held by women in Japan. This is very fully described in Chamberlain's *Things Japanese*, published as late as 1890, and his narration is fully confirmed by my own observations some years earlier. He says: 'A woman's lot is summed up in what are termed the three obediences - obedience while yet unmarried to a father; obedience, when married, to husband and the husband's parents; obedience, when widowed, to a son. At the present moment the greatest duchess or marchioness in the land is still her husband's drudge. She fetches and carries for him, bows down humbly in the hall when he sallies forth, waits upon him at meals, may be divorced at his pleasure. Most Japanese men in this very year of grace 1890 make no secret of their disdain for the female sex. The following seven reasons for divorce are given from the book of a native philosopher,[1] entitled *The Whole Duty of Women*:
1) A woman shall be divorced for disobedience to her father-in-law or mother-in-law.

2) A woman shall be divorced if she fails to bear children, the reason for this rule being that women are sought in marriage for the purpose of giving men posterity. There is however no just reason for a man to divorce a barren wife if he has children by a concubine.

3) Lewdness is a reason for divorce.

4) Jealousy is a reason for divorce.

5) Any foul disease is a reason for divorce.

6) A woman shall be divorced who by talking over-much and prattling brings trouble on her household.

7) A woman shall be divorced who is addicted to stealing.

Add to this that 'She must sew her father-in-law's and mother-in-law's garments, and make ready their food. Ever ready to the requirements of her husband, she must fold his clothes, and dust his rug, rear his children, wash what is dirty, be constantly in the midst of her household, and never go abroad but of necessity', and the status of women in Japan becomes pretty clear.

Mr Chamberlain gives, in footnote, the opinion of some of the *literati* of Japan delivered in 1888. 'The subordination of women to men is an extremely correct custom. To think the contrary is to harbour European prejudices. For the man to take precedence over the woman is the grand law of heaven and earth. To ignore this, and to talk of the contrary is as barbarous as absurd.'

Fully alive to the position and the circumstances of women in Japan (my views and opinions being those set forth in Mr Chamberlain's *Things Japanese*) I approached with some apprehension the task given me of selecting English women. Consulting the members of the Embassy, I found that their desire was for a merely mechanical education. They had no wish that any attempt should be made to improve the morale, or in any way to alter the recognised position of their women. At their request, therefore, I gave up the idea of seeking for refinements or high attainments in those to be selected, but to obtain persons who could impart a good knowledge of ordinary household duties.

A lady who had lived in Japan with her husband, while I was there, but who was then widowed, and not well off, seemed to my view to be a most likely candidate for the position of matron.

Having mentioned the matter to Sir Harry Parkes, he took a view distinctly opposed to that of the members of the Embassy. He thought it might rather tend to degrade English women in the eyes of their Japanese sisters, if half-educated representatives were sent among them; and he recommended a Miss B. whom he described as a highly refined person, and who was willing to accept the position.

On the other hand, I feared the introduction of delicate refinement into a state of society, the moral and ethical standards of which were so completely opposed to those of Europe; but I deferred to the views of Sir Harry Parkes to the extent of laying them before Ito. As the Embassy was at this time in Paris, I did so by letter. Having sent a copy of this letter to Sir Harry Parkes, I received the following:

'57 Westbourne Terrace
23rd December 1872

My dear Mr Brunton,

I am much obliged for your letter and the copy of your letter to Ito, which, I think, states the case very clearly and fairly. I do not however share your view as to the disqualification of a person of superior feeling and refinement for such a post. I suppose the tone of Japanese ladies is much the same as that of the Chinese, and there I have known the force of example and principle of the European lady check irregularities or lightness of demeanour and ensure the observance of strict decorum in schools and assemblies of young native women. Do you not think the self-respect and dignity of a true lady is more likely than any other quality to prevent impertinence, or the taking of liberties with her? I do not advocate any pressure in the way of education, moral or otherwise, but I do think that the great force of example is the most potent agent you can employ to sway and form minds, although immediate results are not to be looked for. It appears to me that the influence of a painstaking and conscientious lady, one who is resolved to devote herself to her special work, and give her pupils or learners the consideration which their position calls for, so obtaining their respect, must be for good; and the opposite will result if she be not respected.

Whether obtaining the services of superior persons grows out of this proposal for seamstresses, or be treated apart from it remains to be seen. Japanese women will soon feel the need of such instruction, and I think they would be better provided in England than if they were to look towards the United States.

If Ito approves the change of plan which you have forwarded to him, I should think there would be an advantage in having two second-assistants. I feel satisfied that if such an establishment were at all judiciously conducted and rendered attractive by its superior as well as its useful tone to Japanese women, Miss B. would find it difficult to keep pace with the demands upon her time, and should desire no higher object than to feel she was aiding in elevating the women of Japan.

Yours very truly
Harry S. Parkes'

While appreciating the force of Sir Harry Parkes' remarks, a consideration presented itself to me which prevented my bringing any pressure on the Embassy to further Sir Harry Parkes' views. Miss B. whom he desired for the position, was a friend of Lady Parkes' family, who were extensively engaged in missionary work in China.[1] Christianity being at that time forbidden by edict[2] in Japan,[2] I feared that attempts at proselytising might be made by this lady, and my introduction of her would be regarded as an act of disloyalty to my employers.

Mrs M. on the other hand, from her residence in the country, had a personal knowledge of the peculiarities of the people, and like most foreign residents had no sympathy with the missionaries; and, further, from her married experience, could efficiently instruct the Japanese women in those household duties which it was desired they should learn.

All hope of Miss B. being appointed was dissipated by the following letter:

'No. 10 Rue de Prestaururz
Paris, 2nd January 1873

Dear Mr Brunton,

I received your note of the 23rd of December last relating to the female school at Yedo. I am still in favour of my first plan which I arranged with you before leaving England.

Sir Harry wrote to me a few days ago and I have answered his letter giving the same opinion as I now do.

I will stay in Paris two weeks more. Have you engaged all the engineers and artisans for shipyards?

I am yours,
Ito.'

On the same day the following was received:

'57 Westbourne Terrace
1st January 1873

My Dear Mr Brunton,

I have received the enclosed note from Ito, and yours

130

almost simultaneously. It is dated the 28th of December but the postmark shows that it was only posted in Paris last night.

I think it shows that Ito does not wish to engage a lady of the quality of Miss B. Of course you can only carry out his wishes. His plan seems to be a very imperfect one, but they can certainly claim to be allowed to please themselves.

The last novelty that has reached me is the proposal to engage Japanese seamstresses to teach embroidery &c. in England.

Yours very truly,
Harry S. Parkes.'

This leaving the way clear for the lady I had first in view, I deemed it prudent that she should place herself in communication with Sir Harry Parkes, with the result that I received the following letter:

'57 Westbourne Terrace
10th January 1873

My Dear Mr Brunton,

This evening I have received a note from Mrs M. saying that she wishes to take my opinion on the appointment proposed to her. It would be impossible for me to see her tomorrow, for I am engaged from 9 a.m. to 11 p.m.; but this apart, I really don't see in what respect my opinion could be of any service to her. I don't know what to give her an opinion upon. I can form no idea of the kind of tuition she will be required to give, nor the persons she will be called upon to instruct. You, on the other hand, do know what Ito's objects are, and can describe these to Mrs M.

I have an opinion as to what the character of the instructor of Japanese ladies of official families should be, but you know that this view is not concurred in, or is not thought necessary in the present case, by Ito.

I feel therefore that I am entirely incompetent to give Mrs M. advice, nor can I accept any responsibility in the matter of her appointment. If she is to come into contact with native society of the questionable character you spoke of, she, and she only, can decide whether she should accept, and if she accepts, how she should deal with the position.

The information which you can supply will I think be much more useful to her than anything I can tell her. Would you kindly explain this to Mrs M. or show her this note. As I am very much pressed, she will excuse me not repeating it direct to herself.

Yours very truly,
Harry S. Parkes.'

Being still further pressed to give his opinion as to whether the appointment was one which a lady in Mrs M.'s position could safely accept, Sir Harry Parkes wrote the following further letter to the author:

'57 Westbourne Terrace
Hyde Park West
12th January 1873

Dear Mr Brunton,
My position in regard to Mrs M.'s appointment is simply one of entire neutrality, I can neither approve or disapprove it and with such information as I, at present, possess, I can form no opinion as to the scope of the duties, or the comforts, or the respectability of the position. All that I gather from Ito's letter is that he is determined it shall bear an inferior stamp, and that the social position of the matron is not to be that of the lady. In this, I think he is mistaken, and Mrs M., it appears to me, should be prepared for much unpleasantness, and for all the inconvenience contingent upon what she may find to be a false position.

These being my views, I felt I could do Mrs M. little service in the way of an adviser. From you she receives her information as to the duties she is undertaking and the position she is to occupy, and it then rests, it appears to me, entirely with herself (or one of her family) to weigh the disadvantages as well as the advantages of the situation proposed to her and to come to a determination in the matter.

I wish you to show this note to Mrs M., that both she and her brother, who has written to me about her, may clearly understand that, while I wish her all success in her work, I can accept no responsibility in regard to the conditions under which she goes out.

Yours truly,
Harry S. Parkes.'

The lady referred to and her friends, having given the matter full consideration, it was eventually decided that she should accept the appointment, and with her assistance and approval, the two others being appointed, the three travelled by the same steamer which took my wife and myself to Japan. Having obtained quarters in Tokio, and the necessary accommodation for their pupils, the ladies proceeded to impart to young Japanese girls a variety of information for a considerable period, but the experiment cannot be regarded as a success. The independence of their position and their utter change of circumstances led to disputes and quarrels arising among themselves; the matron eventually became invalided. One of the others had the imprudence to marry a missionary, while the third entered an English family as a companion.

Thus this particular attempt at elevating the Japanese woman failed, and so far as I know no further effort in that direction has been made.[3]

The Embassy having left London for Paris on the 15th of December, 1872, and I having executed all the commissions entrusted to me by its different members, I left on my return to Japan on the 12th of February, 1873, arriving in Yokohama on the 4th of April.[3]

133

Sir Harry Parkes had arrived back some time before this. Having completely failed in their political design, after nearly two years absence, the Embassy returned to Japan on the 13th of September, 1873, its members being in an exceedingly dissatisfied and reactionary frame of mind.

The ambition of the Japanese rulers had long been to obtain jurisdiction over aliens. Refusal by the Treaty Powers seemed to them not only a slur on their civilisation but also a humiliating personal stigma. Little wonder, therefore, that their association with foreigners became strained, their dealings with them less frequent, and their readiness to be influenced by them almost destroyed.

There is no doubt that in 1872, neither the Government nor the social system of the Japanese were in a state to warrant their being entrusted with the judicial control of people from the civilised nations of the West. Japan had just emerged from a regime of feudal rule, in which the sword was almost the only means of arbitrament between parties. There was no public written law. Witnesses as well as accused were subjected to torture and horrible cruelty. The Tokio Government had taken steps to obtain from some French lawyers, Messrs Boissonnade[4] and others whom they had employed, a code of laws based on the Code Napoleon. It was argued, however, by foreigners that the possession of a code, however excellent it might be, did not guarantee its efficient administration. Hence the Japanese request for jurisdiction in 1872 was firmly refused, though in January 1899 they were granted the long desired concession by all the Treaty Powers.[4]

CHAPTER THIRTY-ONE

THE JAPANESE IN BAD TEMPER

THE SPIRIT of sullen anger, bred by chagrin and disappointment, through the failure of the political objects of the Embassy, led to the adopting of a distant and overbearing demeanour towards foreigners. The policy of a strict maintenance of their rights, a desire as far as possible to do without the assistance of their foreign servants, a determination to keep those whom they required in a wholly subordinate position, were distinguishing features in the attitude of the members of the Government at the end of 1874.[1] An almost entire disregard for the advice of foreign helpers was ostentatiously shown. As a direct result, wholesale dismissals of European employees took place. Works of improvement were abandoned. The influence and advice of the British Minister, once so powerful and so highly valued, was ignored.

The marvellous exhibitions of power, invention, resource, and the results of industry which had been presented to their virgin minds in the United States, and in every country in Europe, seemed wholly lost. Their sense of intense dissatisfaction, arising from the fact of their being not yet supreme in their own country, overcame every other consideration. This spirit retarded in the most serious manner the expansion which had begun so favourably.

The functionary in charge of the lighthouses, carrying out the instructions of the Government, proceeded to do his best to thrust the European staff into positions of mere drudges without authority or influence.

Taking action independently and unknown to me,

making direct communication with the subordinate Europeans, and giving instructions with a peremptoriness which was intended to brook no contradiction, he assumed an attitude at once haughty and autocratic. His communications were all made in writing, and all documents were required to bear his seal before being regarded as authorised.

This line of conduct was evidently instigated by his superiors in Tokio, as various notifications bearing on the relations of the Government with foreigners were at the same time issued. Presumably intended to define more accurately the position of aliens, some of these were exceedingly quaint in their character.

The Government at this time was somewhat agitated over financial matters,[1] and determining to retrench as much as possible, issued the following notification:

'Council of State, Tokio
12th August 1874

It is hereby notified that owing to the great expenditure, which has become necessary, in the many public offices, special economy must at the present time be exercised. The different Departments of Public Works are to discontinue all new forms of expenditure - this not applying to buildings already commenced. New advances to persons with the object of assisting them in their businesses are to be discontinued, excepting those of an absolutely necessary kind.

Any balances in hand left over from former years are to be paid over to the Finance Department.

All Public Offices are to make investigations in order to discover methods for further economy, and to report the results without delay.

Sanjo,
Prime Minister'

This notification, though issued by the Prime Minister, was not very fully acted up to, and it is believed that just at the period of its publication considerable difference had arisen between members of the Government in connection with financial affairs.[2]

CHAPTER THIRTY-TWO

THE YOKOHAMA HARBOUR SCHEME

ABOUT THIS TIME Okuma, Minister of Finance, together with Ito, Minister of Public Works, was exceedingly desirous of having constructed in Yokohama Bay a pier or wharf, alongside which the rapidly increasing shipping visiting the port could lie and load or unload cargoes.[1]

Receiving instructions in August 1873 to prepare correct plans and estimates for this, I was, during that and the following year, much occupied with this work. The scheme gradually expanded until it became one of considerable magnitude, embracing as it did the construction of two concrete piers, each one a mile long, enclosing five hundred acres of water, and estimated to cost $2,500,000. Completing in full detail the plans, which received the approval of both Okuma and Ito, and being informed by the former that the necessary funds for their execution had already been provided, I was somewhat dismayed to learn that the majority of the Japanese Ministers adopted the Premier's view of the financial situation, as embodied in his notification, and that they declined to sanction its construction.

The following from Sir Harry Parkes explains to some extent the situation:

'10th February

My Dear Mr Brunton,
The pier scheme is certainly attaining considerable dimensions. Okuma told me on Saturday that the larger plan was in contemplation, and that the necessary money would be found.

Yours very truly,
Harry S. Parkes.'

'9th March

My Dear Mr Brunton,
Thanks for the list. The suspension of the harbour works lends confirmation to the report that Okuma's finance is being attacked. I shall be curious to know the result of your interview with Ito. You will doubtless be careful to get from him what he has to propose about yourself.

Yours very truly,
Harry S. Parkes.'

This question of a harbour for Yokohama excited considerable controversy, both among the natives and the foreign community. A certain section of both objected to it, on the grounds that it involved a large and needless outlay, which might be diverted into channels of greater general advantage.

Anxious that the matter should be clearly understood, I contributed various letters[2] to the local papers, affording information concerning it. Referring to one of these I received the following:

'Monday

My Dear Mr Brunton,
I congratulate you on the very able letter which

appears in the *Mail* of Saturday, on the Harbour scheme. It is full of information and effectually demolishes your opponents. I think you will now find that the plan is understood by the community.

Yours very truly,
Harry S. Parkes.'

Largely debated on all sides, the scheme was staunchly upheld by the two responsible Ministers, Okuma and Ito; but the opposition eventually overcoming them, I received the following from Hayashi, later Minister to Great Britain and Minister of Foreign Affairs in Tokio, but then in the Public Works Department:

'Public Works Department
Tokio, 8th March 1875

Dear Sir,
I am directed by the Minister of Public Works, to inform you that the Government has finally resolved not to carry out the harbour scheme at Yokohama, designed by you, at the present moment.

The Minister (Mr Ito) will be glad to confer with you concerning your future appointment, if you will come to Tokio. You will generally find him at his office in this department any day after two o'clock.

I remain,
Yours truly,
T. Hayashi.

R. H. Brunton Esq.'

The harbour scheme thus dropped out of sight, but not before the Mikado himself had become interested in it, and had examined the plans.

Announcing his intention of visiting the lighthouse establishment at Yokohama in March 1874,[3] the

Emperor expressed his desire not only to be shown the machinery and apparatus connected therewith, but to have the details of the proposed harbour improvement scheme fully explained to him.

Arriving on horseback, with a large retinue and accompanied by the Empress and her suite in carriages and all arrayed gorgeously in silk, the Emperor received first the foreign representatives who were presented by his Minister of Foreign Affairs, after which Ito, Minister of Public Works, presented me and the European staff. A prettily decorated side room having been prepared, he, in this, examined a chart showing the position of all the lighthouses then erected, as also a table giving the cost of each. A bird's-eye view of the proposed harbour had been specially made for the occasion, in which His Majesty showed great interest making also a very close and intelligent examination of the plans.[1] With Hayashi[42] acting as interpreter, he accompanied the Lighthouse Commissioner and myself through the various machine shops, stores and other buildings which formed the establishment. He scrutinised carefully the lamps and other apparatus by which light is made to pierce great distances, and expressed himself as much gratified by the information afforded him.

On the following day, the Establishment was thrown open to the general public, and visited by thousands of Japanese who all came dressed in their ceremonial garb.

CHAPTER THIRTY-THREE

MAINTAINING DISCIPLINE

THE EFFORTS for the effacement of the Europeans in Government employ were determinedly persisted in, and to his credit be it said, the Commissioner engaged in the task with courage and energy. Strongly feeling, however, that the work I was engaged on was of an international character, involving as it did the safety of foreign vessels, I resolved to make, at whatever personal sacrifice, the most determined stand against the indignities attempted upon me and my staff. Communications between the Commissioner and myself thus partook of the character of one long continued struggle, maintained with varying success on either side. Lucky in having a man at the head of the Public Works Department (Ito) so enlightened and so favourably impressed towards me personally, I succeeded eventually in accomplishing the erection of many lighthouses, and in placing the whole system in a well organised condition, without an actual rupture of good relations; but success was only won amidst a mass of heated contention.

In the construction of new buildings, an absolute refusal was given to the employment of any European artisan as inspector; and in some cases specifications describing the work to be executed were without consultation or notice substituted for those supplied by me. The work done in this way was characterised by the most absurd non-conformity to accepted building principles, and only on my urgent representations were the condemned portions, when possible, taken out and rebuilt. It was not uncommon to find men beating up into a powder cement which had already been used and lost all its adhesive properties for a second application. I noticed bonds between adjacent portions of buildings completely neglected; courses of brickwork laid out of level, the deficiency being made up with thin pieces. Numerous other similar errors, which might naturally be expected from persons having no experience in the work they were attempting to execute, were common.

In reply to my request for the employment of experienced superintendents, the contention of the Commissioner invariably was, that it was my own and my assistants' duty to teach the Japanese, and therefore, artisan inspectors were superfluous.

Having pointed out to the Commissioner that civil engineers were supposed to have only a general knowledge of building operations and were not practical mechanics, I contended that they were not, therefore, well fitted to educate ignorant workmen. I urged further that it was necessary to have a resident inspector at each structure in hand, and that the duties the engineers were engaged to perform were of an entirely different nature.

By discursive arguments of this nature, the Government functionaries were at length compelled, if not completely, at least to some extent, to meet my views, and nearly all the work was carried out under European supervision.

Possibly a more important matter, involving as it did the safety of mariners, was the employment of trustworthy men to tend the important ocean lights. A voluminous book of instructions, compiled from British, French and American sources giving with great detail the duties required from men in charge of a light, was drawn up and translated into Japanese. A three-storied model lighthouse - a brick building twenty feet square and forty feet high - was built in Yokohama, and one of the most skilful English lightkeepers was appointed to instruct the Japanese in the manipulation of lamps and apparatus. The space occupied by workshops and store rooms, covering about four acres of ground, was on the shore. The yard was fronted with a stone jetty, outside of which boats could lie. There was also a crane and the lines of rail ran from it to the storehouses. Here the iron buoys were made and the wooden lighthouses were fitted previous to their being sent to their destination. The whole work was carried on by the help of two assistant engineers, Mr James Ritchie and Mr S. Fisher. We had also twenty-four European artisans, inspectors, lightkeepers, etc. Hundreds of native young men were drafted into the service to pick and choose from, yet it was found impossible to obtain an adequate supply of really trustworthy native lightkeepers.

In those days, before there was a public school system, before a systematic civil service, or the training of the army and navy, all of which have inured the Japanese to modern discipline, there appeared to be ingrained in the nature of the applicants for salaried positions an apathetic indifference to routine duty. Their old habits did not permit of their giving that punctual and systematic attention to routine, which is so necessary in a lighthouse. This feature extended also to the higher officials. They did not regard departures from a line of duty, or an irregularity of conduct in the lightkeepers with a sense of seriousness necessary to ensure non-repetition. The discipline so indispensable in such a service was not maintained, nor was the punishment, ordained by the regulations for breach of duty, duly inflicted. Quite the contrary. On my making any complaint of irregularity, the functionaries invariably showed a disposition to screen or excuse the delinquent.

Sleeping on their watches; allowing their stations to get into a state of disorder and slovenliness; threatening European lightkeepers with their swords;

143

in many cases letting the lights become extinguished; allowing machines of revolving lights to stop, thereby entirely destroying their distinctive character; drunkenness; falsification of returns and attempts at barefaced deception, were some of the delinquencies which were brought home with alarming frequency to Japanese lightkeepers. For these offences they seldom or never received such punishment as would be regarded adequate in Western countries, and as was prescribed in the book of regulations, which had been formally approved. Silly excuses for irregularities were often made. On one occasion, it being discovered that a man had deliberately left the lighthouse during the hours he should have been on watch, he averred that he was carefully watching the light from his dwelling house.

In 1874 I pointed out officially that thirty-seven efficient lightkeepers were then required for different stations, but that out of one hundred Japanese tried, only twenty could pass muster, while nine only out of the twenty were known to be efficiently trustworthy to warrant their being placed in charge of a station without a skilled foreigner over them.

Strongly urging the appointment of other Europeans, I was met by the most determined resistance. I therefore felt it necessary to put on record my protests against my advice being ignored, and to lay the facts, and the general state of disorganisation of the lighthouse establishment before Sir Harry Parkes.

Hints also reached the British Minister's ears of a proposed wholesale reduction in the number of the foreign employees, and that no renewal of agreements with them would be made. The proper maintenance of the lights under such circumstances became a matter of some anxiety to him, as the following letter shows:

My dear Mr Brunton,

I have written officially to the Government, pointing out that the reduction in the foreign staff should be gradual, that the present staff of mechanics and lightkeepers is below that stated in the memorandum supplied me (six of the former, twelve of the latter) that at least this number is requisite, and that they should be engaged on longer terms, so as to secure good men. I have also pointed out that an experienced Chief Engineer is the last man that can be dispensed with safely, but that the power of the present chief (yourself) and of the lightkeepers now engaged, to render good service, is rendered almost null by the system pursued by the Japanese Government, of depriving them of all authority. The authority of the Chief Engineer, I have said, should be supreme over every one, whether Japanese or foreigner, employed in any administrative branch of the service.

The letter has rather surprised the Foreign Minister, and I believe Mr Ito is to see me on the subject. The correspondence will, of course, go home.

I have mislaid the list of foreigners in lighthouse employ, or rather have given it to Mr Aston,[1] who is away on short leave and I cannot get at it. Could you give me another copy, and pardon the trouble I occasion you?

You supplied me last year with copies of three letters you wrote to Mr Ito and Mr Sato upon lighthouse construction and the supply of stores. Did you ever receive any answers, or what was done about them?

Are you writing now another report on your late inspection?

Your last report was August or September 1874 when you had been obliged to return without inspecting all the lights. I should like to have your report of *Yebosi* (iron) lighthouses which they insisted upon building without your assistance.

I shall very gladly give you a testimonial; but do you wish this at once, or later on, when your departure is determined on? I am scarcely yet in a position to speak of your leaving the Japanese service, and my testimonial must be a more serious document than that of Captain St John,[2] and must be sent home.

I am,
Yours very truly,
Harry S. Parkes.'

In reference to Sir Harry Parkes' inquiry regarding construction, it may be mentioned that most of the lighthouses were built under direct European supervision. A few were erected by Japanese alone, these being probably erected twice over before being completed, owing to the mistakes of the natives, and their showing signs of the most faulty workmanship. In regard to the supply of stores, after years of contention, during which time many urgent messages were received reporting shortness of oil from one lighthouse, wicks from another, lamp-glasses from a third, and so on, the commissioner announced that he had at length come round to my way of thinking, and had determined that one year's stores should be rigorously maintained at every lighthouse.

It was in this way that the lighthouses were saved from the slough in which other works undertaken in Japan were at this time floundering, and were regarded by people of all nationalities with satisfaction.

The following letter was received from the United States Consul at Yokohama, Col C. O. Shepard:[1]

'United States Consulate
Yokohama
2nd September 1873

R. Henry Brunton, Esq.

My dear sir,

To my mind, no foreign improvement adopted by the Japanese has shown the same valuable results as the Lighthouse Department, and in my annual report to my Government, now in hand, I desire to speak at length upon it. Its management is admirable and its benefits great, and if you will give me any facts or figures of interest, I should be greatly obliged.

Yours very truly,
C. O. Shepard.'

This gentleman having sent to me a copy of his report, I transmitted an extract from it to Sir Harry Parkes, and received the following in reply:

'Yokohama
Friday

My dear Mr Brunton,

Thank you very much for sending me the extract from Mr Shepard's report. I am very glad to see he has formed so just an estimate of your work. Do you know the date of his report, and when it is likely to be published?

Yours truly,
Harry S. Parkes

P.S. I heard yesterday from Vienna[3] that among the various prizes awarded to the Japanese at the Exhibition there, one was a Diploma of Honour awarded to the Lighthouse Department.

H.S.P.'

CHAPTER THIRTY-FOUR

KEEPING UP THE STANDARD

KEEPING MYSELF in constant communication both with the captains of the different warships and the commanders of mail and other steamers frequenting Japan, I learned much from them concerning the appearance of such lights as they passed. It was very gratifying to find that very general commendation was expressed. That results so satisfactory were obtained, was chiefly due to the pressure which I was enabled, through Sir Harry Parkes, to bring on the servants of the Japanese Government.

According to agreement duly made, I received one year's formal notice, in March 1875, that my services would not be further required. I shall always retain the most grateful remembrance of the consideration shown me by the Japanese Government, when, in accordance with its settled policy, they declared their inability to employ me any longer.

Called to Tokio to an interview with Mr Ito, I was told that the Government fully appreciated the assistance I had given it, at a time when the Emperor's servants were only beginning to open their eyes to the ways of Western civilisation. While it had been decided that I was not to retain the charge of the Lighthouse Service, Ito would find other employment for me, as the Government did not intend to dispense with my services.

However, three days after this interview, the Commissioner informed me that although Ito was prepared to do what he could to obtain other work for me, he had been instructed to give me formal notice of severance of my relations.

Honestly believing that Minister Ito desired to retain my services in some other sphere, nevertheless discerning after the decision to postpone the Yokohama Harbour Works the uncertainty of the Government's finances, I felt that his power to accomplish his purpose was doubtful.

This anticipation proved to be correct. My engagement with the Japanese Government terminated in March 1876, when I left the country and returned home, after having been the recipient of many presents, and entertained at many farewell banquets.

The following is Ito's written appreciation[1] of my services:

'Public Works Department
Tokio, 4th April 1875

Sir,

I have much pleasure in informing you that during the time you have been in the service as Chief Engineer of the Lighthouse Department, the work entrusted to you of erecting and maintaining lighthouses on the coasts of Japan, has been executed to the most complete and perfect satisfaction of the Government. These lighthouses now number above thirty, and both as regards their design, their construction and working, there has been every reason to be satisfied with them. Before parting with you, at the end of your long service, I beg to express the hope, and indeed I cannot doubt that, though the future policy and plans of the Government do not require your further services, your ability, skill and knowledge, will ensure you a future career as successful as that in the service of this Government.

I am, Sir
Yours faithfully,
Hirobumi Ito
Minister of Public Works

R. H. Brunton Esq. Chief Engineer.'

149

The following was received from the Foreign Office in London:

'Foreign Office
10th January 1877

Sir,

I am directed by the Earl of Derby to acknowledge receipt of your letter of the 8th instant, requesting, with a view to the furtherance of your prospects, to be furnished with the opinion of Her Majesty's Minister at Yedo upon the work done by you in connection with the establishment of the Japanese Lighthouse System, and I am to inform you in reply, that his Lordship gathers from Sir Harry Parkes' despatches, that your labours in Japan were satisfactorily performed, and are appreciated by him, and that the arrangements made by you have worked well, with every prospect of continuing to do so.

I am,
Sir,
Your obedient servant
Tenterden.'

On the completion of my term of service, the Lighthouse Department had established, or had in hand, thirty-seven ocean lights, nine harbour lights, three light-vessels, fifteen buoys and eight beacons.

The total expenditure on these has been given by the Government officials as not far short of one million sterling, but, as has been already explained, I was unable to keep any record of costs, hence I have no means of checking this figure.

CHAPTER THIRTY-FIVE

THE RIU KIU ISLANDS

WITH HIS USUAL consideration, the British Minister decided to accompany me on my last trip round the coasts of Japan in the *Thabor*.

Landing at every lighthouse and examining the system which I had devised for their maintenance, Sir Harry Parkes expressed himself as much pleased in what he had seen, and on the whole satisfied, though defects which might have been avoided were in a few instances observable.

Calling at various places along the coast, not formerly visited and seeing many new and interesting features of the country, it was decided at Sir Harry Parkes' request, that the ship should also go to the Riu Kiu Islands.[1]

These form a portion of a chain of islands which extends from the most Southern point of Kiushiu, Satanomisaki, to the north of Formosa.

At that time the Chinese claimed this island kingdom as a vassal state, but the Japanese were asserting their undoubted historic rights. The 'kinglet' Shotai had long enjoyed or suffered the disadvantages of a double protectorate, paying tribute to both China and Japan. In 1879 the Japanese Government took complete possession of the islands, and removed the 'king', Shotai, to Tokio, where he was made a marquis.

The seat of Government, at which Sir Harry Parkes desired to call, was then on the main island, or Okinawa, from which the present prefecture takes its name. It lies about three hundred miles from the nearest point of the main island of Japan.

The capital, named Shuri, is situated about three miles inland from the port called Nafa,[2] where the vessel anchored. The harbour is not one of which vessels can make any extensive use, there being only a slight indentation in the coastline with little or no protection from the westward. On all sides are precipitous coral reefs, which being only a few feet below the surface of the sea, create most formidable dangers.

On arrival at Nafa, it at once became evident that the customs of the people were very different from those of the Japanese. Several boats came off to the ship. Instead of the tidy, swift, picturesque craft universally seen in Japan, were canoes propelled by paddles. Similar to those seen in most tropical countries, they were hollowed out each from one log of wood. Several huge junks were seen in a creek leading out of the bay, but instead of being built in the Japanese style, they were Chinese in form and build, and had the traditional huge eye painted on each bow. These made periodical trips to Fuchau,[3] the nearest port in China.

Formed entirely of coral, as evidenced by seeing this material on the tops of the hills as well as in all parts, the island of Okinawa affords an interesting example of a coral formation which has been subjected to volcanic upheaval. Being still liable to violent and frequent earthquakes, it shows that it is not yet far removed from volcanic action.

Trees on the island are small, and the wood used is chiefly imported from Japan. Stone, therefore, enters largely into building operations, and the execution of the mason work is much superior to any seen in Japan. All bridges on roads or elsewhere, are built of stone, the openings being spanned by arches of elliptical form. The streets are paved with blocks of stone, having irregular surfaces, which makes walking over them a somewhat painful operation. It was observed that in front of temples, places of importance, or the dwellings of persons in a high station, the roads were laid with broken stones bound together by clay much in the same way as macadamised roads are. As this gave a perfectly smooth surface, it is to be regretted that the system had not been more largely adopted. The main road from Nafa to the capital Shuri, three to four miles long and thirty to forty feet wide, is laid for the whole distance with these irregular blocks of stone.

A little distance out of Nafa, under some fine old fir trees, are numerous graves of Europeans. Each of these had a huge block of stone work placed over it, on which there were inscriptions engraved. From these it was seen that a number of Catholic missionaries were buried there, four men of the American Squadron, and one man belonging to the *Alceste*.[4] As showing the good feeling existing between the English and the inhabitants, the inscription cut on the stone work over this grave spoke of the memorial having been erected by the 'King and inhabitants of this most hospitable island'.

The people are burdened with the maintenance of a large class of idlers, similar to the old Samurai of Japan, who live upon certain hereditary privileges granted them by the Government. Known by the name of Daimors, they may be seen lounging about in every street. The trading and labouring classes, like those of Japan, are industrious and polite. Their education is of the most limited range, being confined to one book of the Chinese classics. The island, which is about five hundred square miles in area, has a population of 150,000 or about three hundred inhabitants to a square mile.

There being little or no communication with the outside world, only by the most careful and skilful nurturing of the soil were sufficient products for the wants of the population obtained. Sweet potatoes formed the chief food. Patches of rice-fields here and there were visible. Many groves of sugar-cane were observed. Oranges of a peculiar aromatic flavour grew on the island, but fruit of any kind seemed scarce. The sago palm is cultivated in large quantities, and the sides of all the hills, not otherwise occupied, are covered with it. Small quantities of tea and tobacco are grown, and there are some cocoanut trees which do not, however, bear fruit. The climate is of a sufficiently genial character to allow vegetation to be green throughout the year, and admits of two crops being produced annually. On the author's visit in December the thermometer stood at 75° Fahr. in the shade and the sun was sufficiently powerful to necessitate a return to the sun-helmets, umbrellas and summer clothing.

The streets in the towns present a most desolate appearance. On each side is a blank stone wall ten or twelve feet high, with openings through them here and there sufficiently wide to allow access to the houses situated inside. Each house being surrounded by a wall, conveys the impression of being a prison, rather than an ordinary dwelling. The houses of the better classes are similar to those of the ordinary Japanese, built with upright posts of wood, having a raised floor laid with mats, and sliding screens of paper. The poorer classes live in huts of a very primitive character. The walls are made by two sheets of bamboo netting, which contains between them about six inches of straw. These enclose the whole house, an opening about two feet wide being left for the entrance. There is no flooring of any description, but a small mat is generally laid over the ground, on which the inmates sit. A weaving loom was observed in most of the houses, as each person makes his own clothing. A pig is attached to each human dwelling and pork is said to be largely consumed.

There were no shops in Okinawa, but articles for sale were brought to the purchasers' dwellings. There is, however, in each town a market-place, where various commodities were exposed for sale. Here were seen sweet potatoes and pork in large quantities, and also a few fish. Japanese tea is a favourite beverage of the people, and much of this was in the market. There were also observed Satsuma tobacco and bundles of English cotton twist. The stalls were all in the charge of women. The only money used was small copper coin, each of the value of about one hundredth of a penny, but the silver of some of the party who purchased a few articles was not refused. Some of the women in the market were young, but the majority of them were elderly. A custom which appears universal of tattooing the backs of the women's hands was observable; those of the younger ones having a few marks only, while those of the older women were covered from the wrist to the nails.

The town of Nafa lies on a level piece of ground adjoining the seacoast, while the capital, Shuri, is built on a series of small hills. The latter is a straggling, scattered place and it would be difficult to form any estimate of its population.

The royal castle is situated about the centre of Shuri, occupying the summit of an eminence about five hundred feet above the sea. The dwellings which are on the highest point are built in the form of a quadrangle, enclosing a courtyard about seventy yards square. Opposite the entrance is the largest building, in which the king lived. On the left and right were smaller houses, the residence of the court officials, officers, etc., all of the ordinary Japanese style of wooden house.

The king seemed to take but little active part in the government of the country and had not been visible to strangers in years. Sir Harry Parkes made vigorous effort to see him, but his officials protested that he was too ill to see anyone, and though reminded of the possible discourtesy with which they might be treating Her Majesty's representative, they adhered to this

155

attitude.

The inscriptions to be seen on many monumental stones placed in the streets are written in Chinese. Many are quotations from the writings of Confucius and other Chinese classics, while others go so far as to represent the country as part of China. The general appearance of the towns gives evidence of the existence of a close intimacy with China, and the principles of building have been largely obtained from the continent.

On the other hand, the language[5] spoken comes from Japan, not the modern Japanese, but the language believed to have been spoken in Nippon some centuries ago. While a great part of what was said by the people was understood by the Japanese accompanying the author, many words were used which were recognised as once belonging to the old Japanese vernacular, but now obsolete.

There were no Chinese resident in the island, and but four Japanese. The more frequent intercourse with the latter, in recent years, had led to a greater familiarity with Japanese ways, and the use of Japanese produce.

Altogether this diminutive kingdom presented an interesting admixture of Japanese and Chinese customs and ways.

Now merged into the Japanese Empire, the simple inoffensive inhabitants will doubtless have their institutions and customs revolutionised as Japan herself has experienced revolution.[1]

CHAPTER THIRTY-SIX

PERSONAL JUDGEMENTS

HAVING ENABLED my readers to form a judgement on the character of the Japanese by a narration of my personal intercourse with them, it only remains for me to suggest briefly the reasonable conclusions to be deduced from what I have written.

It is only justice to give full credit to a people who have effected so great changes in so short a time. It would be illiberal to withhold admiration from a nation which, emerging from a state of barbaric ignorance, has in the space of forty years become a power in the world. Not only has Japan so conformed to the principles governing more civilised peoples, that the Treaty Powers have agreed to place in the hands of her Government the lives and properties of their subjects, but the material progress of the nation has been considerable.

The President of the United States, Mr William McKinley, in his message in 1900 to Congress, makes the following reference to Japan:

'The closing year has witnessed a decided strengthening of Japan's relations to other states. Her development shows the competence of the Japanese to hold a foremost place among modern peoples. As a factor in promoting the general interests of peace and order, and fair commerce in the Far East, the influence of Japan can hardly be overestimated.

'Accordingly, the respect given to Japan's advice and opinions in regard to affairs in China is daily recorded in the newspapers.'

But after the portrayal of this people's nature given

157

in these pages, it may not be out of place to sound a note of warning against the tendency to unduly and unwisely inflate our conception of them.

It should be remembered that the Japanese themselves did not initiate the transformation of their country. On the contrary, they resisted Occidental influences with all the vigour at their command. It was only a sense of their utter helplessness that impelled them to restrain from opposition to the demands of the civilised world, being afterwards driven onwards by the resistless energy of His Majesty's minister. No picture was presented of the Japanese pleading to the Western community of nations for admittance within its portals or, when it was placed before them, of receiving it with gratitude and welcome. It was foreign cannon that frightened them into its acceptance,[1] and for twenty years after its reception, it was necessary to have foreign troops[2] in the country to protect what were described as the 'foreign devils' from the ferocity of these people.

Japan has thrown a glamour over people having only a passing acquaintance with her, and, it must be admitted, over some who know her well. While the beauty of her scenery and the quaint domestic habits of her people are a source of enthusiasm to casual observers, the facility for adopting change and the cleverness with which methods, strange to them, are imitated, induce a much too exalted judgement to be formed of their capabilities. Again, the rapidity with which the revolution of 1868 was accomplished may, to a large extent, be accounted for by the doctrines of docility and abject obedience to superiors inculcated in the country, and by the theory of the divinity of the Mikado, when they and he had once been won over to foreign ways.

It is mischievous however, to assume on this account that the Japanese have changed their skins, or become in the short space of one generation the equal, in knowledge, experience, sense of honour, or morality, of the cultured peoples of Europe and America.

The palpable absurdity of the contention which so exalts the Japanese in their own eyes hardly requires pointing out. It is not even necessary to take into consideration the black races of Arabia or Africa, the yellow races, or others, such as North American Indians and Egyptians, in order to form a due appreciation of the varying degrees of development in different branches of the human race. The matter can be fully demonstrated by confining ourselves to a comparison of the Japanese of forty years ago with Western peoples.

Assisting the complacent Japanese, there exists among a large class of Europeans not conversant with the people, a most unfortunate tendency to bestow excessive laudation on them. Descriptions of their presumed cleverness, and of their manners and customs which have been published, are grotesque in this exaggerated praise.

These do not come from people who quietly remain at home, but from persons who have visited, and probably remain in the country for a few weeks or months; and in consequence of their utterances impressions wholly erroneous are popularly formed.

Japanese art has so saturated our conception of the beautiful that it enters into nearly all our productions. To ensure the success of a comic opera,[1] a Japanese title should be given it, and Japanese ways travestied – however ridiculously. To make an attractive street poster, the design must follow on Japanese lines, however unnatural they may be. To be considered as having a knowledge of fine art, it is necessary to stimulate an exaggerated appreciation of Japanese work.

It should never be forgotten that the Japanese are Asiatics, with Asiatic characteristics and Asiatic sympathies.

Nevertheless, I see clearly that there is no race in the East with whom an alliance may bring so much commercial advantage to our country and so much moral stamina to the Japanese.

Their insular position in relation to another great continent induces a resemblance in their needs and aims; their temperature and latitude suggests further comparison. The Japanese feel that from no race can they learn so much, or benefit so greatly from association with as the British. But the ways of the past ought to be considered, if one has any desire to draw a horoscope of the Future.

POSTSCRIPT

MR. BRUNTON returning to Great Britain from Japan in 1876, received a Telford premium for his valuable paper on the Japan Lights, printed in Minutes of Proceedings of the Institute of Civil Engineers (vol. XLVII, p. 1, p. 4).

In 1878, he became manager of Young's Paraffine Oil Company of Glasgow. He made profound studies of the somewhat complex and intricate processes connected with the destructive distillation of the Scotch shales, and the refinement of the wonderful products obtained therefrom. In 1881 he presented a paper to the Institution of Civil Engineers on The Production of Paraffin and Paraffin Oils, for which he again obtained a Telford Premium.

In 1881, and during the succeeding fifteen years, in partnership with a young friend, he was engaged in the manufacture of architectural ornamentation. During this time he succeeded in solving some of the difficult problems of acoustics in theatres and public halls. Later he practised in London as an architect and engineer, having charge of the design and erection of some of the largest buildings in Great Britain.

In the few months preceding his decease, he prepared a manuscript, which he entitled 'The Awakening of a Nation: being a description of the entry of Japan into the Sisterhood of Nations, with an Elucidation of the Character of the People from personal experience'. The first part was historical in form. His sources of information were 'the Japan Papers presented to Parliament; [Anglo]-Japanese local newspapers; *Things*

Japanese, by B. H. Chamberlain; *The Japanese in America*, by C. Lanman; *The History of Japan*, by F. O. Adams; and various papers of my [his] own published in the proceedings of different societies and in different magazines.' The main contents of the second part of the manuscript, condensed and annotated, may be found in the preceding pages of this volume.

Mr Brunton was elected an Associate of the Institute of Civil Engineers on the 7th of April, 1868, and was transferred to the list of Members on the 17th of May, 1873. He was made a Fellow of the Geological Society in 1867 and of the Royal Geographical Society in 1871, and a Member of the Society of Arts in 1878.

Mr Brunton's health failing, after some weeks of illness, he died at his home in London, at 45 Courtfield Road S.W., on the 24th of April, 1901.

NOTES

BY WILLIAM ELLIOT GRIFFIS

INTRODUCTORY

(1) See Chapter on the Recent Revolutions in *Japan, The Mikado's Empire*.

CHAPTER TWO

(1) Terashima Munenori, born in Satsuma, 1830, was a scholar in Dutch and English (pupil of Dr G. F. Verbeck) and active in political affairs from the arrival of Perry. He was in England, 1865–67, Minister of Foreign Affairs, an Imperial Councillor, President of the Senate, or Genro-In, Minister to Great Britain 1872–73 and to the United States in 1883. He wrote a pamphlet against the early formation of the Diet and opposed too rapid changes in government. Terashima was one of the ablest of the Mikado's statesmen in the early years of the present era of Meiji, or Enlightened Civilisation, a good specimen of a progressive-conservative.

(2) In 1872 I had the pleasure and amusement of seeing in the old curiosity shop of the ex-Shogun's warehouses at Shizuoka, 'the Saint Helena of Tokugawaism', much of the accumulated material [including the model American locomotive of 1854] which the various envoys who at different times made treaties presented to 'the Emperor of Japan', the sham Government in Yedo. This was five days after saluting the first telegraph poles in the interior of Japan. *The Mikado's Empire*, p. 547, p. 545.

CHAPTER THREE

(1) This was depended upon as the main source of saltpetre, from which gunpowder was made, the nitrates being obtained by leaching. On the Japanese house and architecture, see R. H. Brunton's paper on *Constructive Art in Japan*, Transactions of the Asiatic Society of Japan: II, 64; III, 2, 18–27; E. S. Morse, *Japanese Homes and Their Surroundings*, Boston, 1886; R. A. Cram, *Impressions of Japanese Architecture and the Allied Arts*, New York, 1905.

CHAPTER FOUR

(1) An American gentleman, Mr Pease from San Francisco, arrived in Yokohama in 1867, with an offer to establish gas works in the settlement within twelve months. He asked that one-fourth the capital should be taken in Yokohama, the rest being furnished in America, and that subscribers to the extent of two thousand burners should agree to take the gas. Though a meeting was held at the British consulate, the scheme was not sufficiently supported. For years Yokohama at night had only the old method of illumination, each traveller or person outdoors carrying a lantern, usually of oiled or gaily coloured paper.

CHAPTER FIVE

(1) *To-jin* (Eastern man, Chinese, foreigner) was the vulgar term for a European; *guai-koku jin* foreign country man being the more elegant word.

CHAPTER SIX

(1) In April 1867, Sir Henry Keppell, the British Admiral, very narrowly escaped being drowned while crossing this bar. On 12 April 1867, Rear Admiral Henry H. Bell, U.S.N. (1807-1867), fleet captain on the *Hartford* under Farragut, and for two years commander of the East India squadron was with Lieutenant J. H. Reid and eleven American sailors, out of a crew of thirteen, lost in crossing the Osaka bar. Their burial in the Kobe cemetery in which British and American marched together to the music of H.B.M. iron clad *Ocean* - one of the many occasions on which the civil, military and naval representatives of the two English-speaking nations have united in mutual sympathy - is described in J. R. Black's *Young Japan*, pp. 123-126.

(2) See Notes on Osaka, by Professor James, T.A.S.J. Vol. VII, Part IV, p. 388. (In 1903, at the International Exhibition of Industry, the display of Japanese machinery and manufactures was amazing in its variety. See Appendix to Clement's *Handbook of Modern Japan*.) The improvement of the harbour is now (1906) being carried out on scientific principles according to Brunton's plans and upon a scale commensurate with the importance of the enterprise.

CHAPTER SEVEN

(1) See T.A.S.J. Vol. V and Rein's *Japan*. Niigata in 1903 had 59,576 souls in the population, with increasing coasting trade and production of petroleum, but was still barred to foreign commerce and large vessels.

CHAPTER EIGHT

(1) As Mr Raphael Pumpelly from America did, at the same time, in Yezo. See his *Across America and Asia*, 1870, and Stead's *Japan by the Japanese*, 1904, p. 445.

(2) See Mines of Sado. T.A.S.J. Vol. III, Part II, p. 74, and Rein's *Japan*. In 1903, Sado had a population of 36,983. See Chapter XX on Mining in Stead's *Japan by the Japanese*.

CHAPTER NINE

(1) The conditions *were* exceptional, and the political pressure was far greater even than the economic one. The Japanese were right. There was true statesmanship in doing quickly what was to be done towards breaking up feudalism and bringing in national unity. Like the Romans, Okubo and Ito would conquer the enemy - feudal conservatism - by roads. Hence the haste with which the reformers drove the railway enterprise to completion. See *The Mikado's Empire*, p. 588.

(2) For accounts of the formal opening of this pioneer railway, see Black's *Young Japan*; *The Mikado's Empire*, p. 565; Stead's *Japan by the Japanese*, Chapter XXIII, and index; T.A.S.J., Vol. XXII; the various statistical publications of the Japanese Government; and for the railways in the Protectorate, see *Corea the Hermit Nation*, IIth Edition, 1907.

(1) For over two hundred years the Dutch traders and their scientific men at Deshima, near Nagasaki, fertilised the minds of the Japanese with science through one of the most cultivated languages of Europe. 'Japan has not been without her scientific giants in the days of old'. One of the pupils of these Dutchmen was the famous Ino Chukei born A.D. 1744, who is referred to above. See his biography, (*The Japanese Picard*) by Cargill G. Knott, T.A.S.J. Vol. XV, p. 173-178. 'Ino has sometimes been called the Japanese Newton; but Seki Shinsuke, a famous mathematician, who invented a kind of differential and integral calculus, has perhaps a greater claim to such a high title.' See 'Maps' in Chamberlain's *Things Japanese*. A handsome modern monument to Ino Chukei stands in Uyeno Park, Tokio.

(2) See Lewis, *Educational Conquest of the Far East*. New York 1903 and *Verbeck of Japan*; and *A Maker of the New Orient*, for pictures of educational activity at this time. Brunton's picture of the impatient Japanese of 1870 matches that of the 16,000 Chinese students now in Japan.

(3) Some of the ablest men of Great Britain filled long and honourable careers in this Imperial College of Engineering (Kobu Dai Gakko). Among them were Principal Henry Dyer, Professor W. E. Ayrton, D. H. Marshall, John Milne, Josiah Conder, Frank Brinkley, W. Gray and James M. Dixon. Dr Edward Divers, professor of chemistry, was in Japan twenty-six years, and for four years principal of the Kobu Dai Gakko. Among his pupils were Dr Shimose, inventor of the terribly effective Shimose powder used in the Russo-Japanese War, Dr Takamine, discoverer of taka-diastase, and many of the present Japanese practical chemists and teachers of chemistry. See *Dai Nippon: A Study in National Evolution*, 1904, pp. 3-4.

(4) Author of a delightful book of sporting and other experiences on land and in the waters of Japan, China and Corea, entitled *Wild Coasts of Nippon*, Edinburgh, 1880.

(5) The geological and other land maps and hydrographic charts made by the Japanese Government are now among the best in the world. For geological sketch of the Empire, see *Japan in the Beginning of the 20th Century*, 1903.

CHAPTER ELEVEN

(1) Ito was sent to America in 1869 to study the coinage system of the United States. His report led the Imperial Government to adopt the decimal system. Previous to the adoption of the milled coinage of 1871, there were over eleven hundred forms of money in the various feudal fiefs of Japan. *The Mikado's Empire*, p. 425.

(2) For a description of his personal appearance, see J. R. Black's *Young Japan*, p. 233.

(3) Many of these gentlemen were pupils of Dr Verbeck who began teaching English, science and literature at Nagasaki in 1859. See *Verbeck of Japan*, 1900; and for biography, Lanman *The Leading Men of Japan*, 1883; and Morris, *Makers of Japan*, 1906.

(4) See J. R. Black's *Young Japan*, pp. 306-308. For the coinage, currency and finance of the empire, see Stead's *Japan by the Japanese*, and the annual *Resumé Statistique de L'Empire du Japon*, for 1906.

CHAPTER TWELVE

(1) For an account of this historic and decisive conflagration, see *The Mikado's Empire*, p. 563. See also 'Fires' in Chamberlain's *Things Japanese*.

(2) In 1903, there were in Japan, exclusive of Formosa, 8,725,544 houses. For house registry and the composition of the Family see Professor Hozumi's paper in Stead's *Japan by the Japanese*, pp. 281-305.

CHAPTER THIRTEEN

(1) A full and very interesting list, descriptive and detailed, of the Government ships of war and auxiliary vessels, and of the ships and steamers belonging to the different clans in 1867, just prior to the Restoration, is given in Stead's *Japan by the Japanese*, pp. 122-125. Several of these, such as the *Beagle*, in which Darwin made his voyage of research, were of historic interest.

(2) *The Religions of Japan*, p. 320.

CHAPTER FOURTEEN

(1) *Verbeck of Japan*, p. 242. See Townsend Harris, *First American Envoy in Japan, 1896*.

(2) See articles on Oshima with history, geology, bibliography, etc., by Chamberlain and Hodges in T.A.S.J. Vols. V and XI.

(3) 'Twenty years ago we were no better than Russians', said a Japanese recently, in standing upon his political rights. Several works on the growth of popular freedom and individual rights in Japan have been published in recent years. In the Constitution of 1889, Chapter II, under article XXXII treats of the rights and duties of subjects. See text and commentary by the Marquis Ito.

CHAPTER FIFTEEN

(1) See T.A.S.J. for history.

(2) To Mr T. B. Glover of Nagasaki many Japanese were indebted for acts of kindness, especially in being aided to get to Europe and to an education. As early as 1863, this gentleman started some of the very first students on their way to Europe and to distinguished careers, and among these Ito and Inouye, two of Japan's most famous statesmen.

CHAPTER SIXTEEN

(1) Mr Brunton read a paper before the same body on 14 November 1876, President George Robinson Stephenson in the chair. His paper is No. 1,451, and was published in a pamphlet of 44 pages, with map and diagrams. The author kindly sent me a copy soon after issue from the press.

CHAPTER SEVENTEEN

(1) See T.A.S.J. Mr E. H. House wrote in 1875 a critical narrative and review of 'The Kagoshima Affair'. See also Chapter XI St John's *Wild Coasts of Nippon*; Kinse Shiriaku, p. 35; Adams' *History of Japan*, Chapter XX. Admiral Togo and many men, since prominent statesmen, were behind the batteries of 1863.

(2) Doctor William Willis was one of the noblest of men. Educated at Edinburgh and in London, he was appointed Assistant Interpreter and Surgeon in Japan, 16 November 1861. He was present at the attack on the British Legation in Yedo, 26 June 1862, and on H.B.M. ship *Argus*, at Kagoshima in 1863. He held other offices, but resigned for service with the Japanese. Was afterwards medical adviser at the British Legation in Tokio and in Bangkok. He died about 1890. He was six feet four inches high and stout in proportion. A man of a kindly and generous nature. 'It is to him in the first place that the Japanese owe better principles in the treatment of wounded enemies. He protested strongly against the methods, or no methods he found in practice during the Aizu fighting in 1869, with good results.' Thus writes W. G. Aston in a letter dated 28 October 1906.

(3) Satanomisaki was at first reached by men who rode in a basket suspended on a trolley. In the Japanese Government List of Lighthouses, Lightships, Buoys and Beacons (pp. 51 with fine maps, 1903) the rock is located in N. lat. 31° E. long. 130.40, near Cape Chickahoff, the dioptric light of 6½ candle power being visible at a distance of 20½ nautical miles. It is reached by boat.

CHAPTER EIGHTEEN

(1) An honourable burial of the corpses took place in the cemetery at Ikegami, near Tokio.

CHAPTER NINETEEN

(1) A considerable body of ephemeral art and literature in the 'Yokohama dialect' such as the *Japan Punch, Flights Outside Paradise, The Far East*, (J. R. Black) and numerous papers and volumes of correspondence reflect the social conditions of this famous settlement in the seventies.

CHAPTER TWENTY

(1) Tomomi Iwakura, born in 1835 in Kioto, was one of the first of the Court nobles to send his three sons for education, first to Nagasaki, as pupils to Dr Verbeck, and then to America. See Lanman's *Leading Men of Japan*; and Morris, *Makers of Japan*. His son, educated at Oxford, is Court Chamberlain to the Emperor.

CHAPTER TWENTY-ONE

(1) Yokosuka was the home of, and part of its ground in the old village of Hemi nearby was the feudal fief granted to Will Adams, the English pilot of a Dutch fleet which sailed in 1598. In the ship *Charity* he reached Kiushiu in 1600. In favour with the Shogun Iyeyasu, he built ships and aided the Yedo Government as one of the first *yatoi*, or hired aliens, in Japan. He remained in the country until his death 6 May 1620. Anjin Cho (Pilot Street) in Tokio and an annual celebration by Japanese still commemorate him. The monuments to him and his Japanese wife stand on the hill overlooking Goldsborough Inlet. This tomb was discovered by Mr J. Walter in 1872. See *Hildreth's Japan*, edited by Clement, 1906, and *The Mikado's Empire*, p. 262. At the Yokosuka dockyard many warships famous in the victories of 1904, at the Yalu, and in the Sea of Japan, in 1905, were constructed. On the 15th of November, 1906, the steel battleship *Satsuma* of 19,000 tons, the largest war vessel in the world, and wholly the product of native Japanese material and labour, was launched.

CHAPTER TWENTY-TWO

(1) See the account of Mr E. H. House, the famous American war correspondent, who accompanied the expedition, in his book *The Japanese Expedition to Formosa* (p. 231), Tokio, 1875, written under the patronage of Count Okuma; *The Island of Formosa* by James W. Davidson, 1903; and the Japanese Government's statistical publication, *The Progress of Taiwan (Formosa) for Ten Years, 1895-1904* (pp. 79) 1905, showing in map, diagrams and tables the astonishing results wrought during a decade of Japanese occupation.

CHAPTER TWENTY-THREE

(1) Mr Brunton's hearty praise of Tsunetami Sano is well deserved. This man was a typical new Japanese, who penetrated the secrets of Occidental power in the modern world. Born in Saga in 1823, Sano went to Nagasaki and under the Dutchmen there learned their language and medicine. He was thus, like nearly all the makers of new Japan, rich in Dutch or European culture. In 1862 he was sent to France to the Paris Exposition, and in Holland superintended the building of a man-of-war. Returning he was connected with the Navy Department. In 1873, as Japanese envoy in Austria and Vice President of the Japanese commission to the Vienna Exposition he remained in the Austrian capital two years. Returning to Japan he was made Senator. In 1880 he became Minister of Finance. He was a passionate lover of art and founder of the Society of Fine Arts as well as of the Philanthropic Society.

(2) This is a mistake. Dr Willis, the British surgeon, had already in 1868 at Kioto, insisted on the Japanese treating the wounded of both sides with equal humanity. In their later campaigns, instead of the old custom of decapitating their wounded enemies or compelling hara-kiri, the Japanese cared for them.

(3) This official stupidity wrought vastly more harm than in simply hampering conscientious servants of the Mikado. More exactly, the inherited mediaeval conservatism of wooden-headed men, who hated both the spirit and the works of the new age, even while drawing salary from the public purse, was at bottom responsible for most of the assassinations of progressive Japanese statesmen. As soon as a cabinet minister showed unusual signs of a vigour that compelled his assistants and all in his Department to strenuous industry and quick adoption of Occidental energy, his life was in danger. Of course other pretexts, often contemptible and usually false, were made use of by the murderers, whose habit was to redress all grievances by the sword. Often these bullies threatened even the lighthouse keepers, as Mr Brunton has shown.

CHAPTER TWENTY-FIVE

(1) Since the recent revival of the shipbuilder's art in Japan, the Japanese have built their own ships, for war and peace, restoring also the ancient and beautiful launching customs. Instead of the traditional breaking of a bottle of wine or water over the bow by a virgin, as the prow of the ship touches the water, a cage of doves is released. This symbolism of freed life finds its basis in a pretty incident in the native mythology.

CHAPTER TWENTY-SIX

(1) I had the honour of audience with the Emperor on this same occasion and remember meeting Mr Brunton in the castle hall in Tokio.

(2) The Hon. William H. Seward, ex-senator and Secretary of State, who had negotiated a treaty with two of the Shogun's envoys in Washington, had a long public and also a private audience of Mutsuhito, Emperor of Japan on 7 October 1870. See W. H. Seward's *Round the World*, New York, 1876.

(3) See Reports of Horace Capron and His Assistants, and the various reports of Professor Benjamin Smith Lyman.

(4) People with straight eyes, full beard and moustache, whose language is of Aryan stock. They are an old 'white race' and form much of the basic stock of the Japanese composite. They once occupied most of the main island of Japan. See the various works, including an Ainu Grammar and Dictionary, mythology, folklore, etc., by the Rev. John Batchelor, for twenty-five years a missionary among them.

(5) Adams' *History of Japan*, pp. 289-292.

(6) 'And presented with a cheque for £500 in appreciation of his services'. See *The Biographer*, London, May 1898, pp. 60-65.

CHAPTER TWENTY-SEVEN

(1) The genesis and original purpose of this embassy, which powerfully influenced the whole nation in favour of modern civilisation, both at the time and through the later voluminous private and official publications of its members, are given in *Verbeck of Japan*, Chapter XIII. Dr Verbeck first suggested its details in a paper to Count Okuma, 11 June 1869.

(2) With the embassy came five young girls, largely through the personal influence of General Kuroda, to be educated in the United States: the Misses Rio Yoshimasu, Tei Uyeda, Sutematsu Yamagawa, Shige Nagai and Ume Tsuda, all daughters of Samurai gentlemen connected with the Government. The third is (1906) the wife of Marshal Oyama, the fourth of Rear-Admiral Uriu, and the fifth is one of the most influential educators in Japan. See Lanman *The Japanese in America*, p. 47.

(3) For their biographies see Lanman and Morris, and life of Okubo by Maurice Courant, Paris, 1904. Most of the secretaries and attachés rose subsequently to high office.

(4) This speech and a complete list of the members of the Embassy are given in Lanman's *The Japanese in America*, New York, 1872.

(5) That has ever been the goal of the new Japan, and to this they are striving - to be 'second to none'. When however, in 1858, Mr Harris, American Consul-General, who made the initial treaty of residence and commerce, proposed consular jurisdiction in an extra-territoriality clause, the Shogun's agents agreed to it without demur. See Townsend Harris' *First American Envoy in Japan*, p. 124.

CHAPTER TWENTY-NINE

(1) See p. 439-440 in Stead's *Japan by the Japanese*. About 60 million gallons of petroleum are consumed annually in Japan. The crude product yields 50 per cent of burning oil, more like Russian or California than Pennsylvania oil. Over 40 million gallons were produced in 1903, worth over $1,200,000. Two companies, the Japan and the Standard Oil, control most of the output.

(1) The noble record of 'Lady Parkes' family' here referred to is given in outline in the Life of Sir Harry Parkes, 1894. Gutzlaff the 'apostle to China', had for his pupils whom he inspired for service, two lads, one of whom became Great Britain's Minister to Japan and China, and the other one of the real makers of the new Japan. They were Harry Parkes and Guido Verbeck. See *Verbeck of Japan*, pp. 142, 143.

(2) Even while this correspondence was going on, orders were being sent to Tokio from the Embassy, that the edicts should be removed and toleration of religion granted. *The Mikado's Empire*, p. 359, and *Verbeck of Japan*, pp. 96, 264-266. Okubo and Ito had been sent back from Washington to urge upon the Imperial Cabinet, among other things, the following: 'Although the Japanese criminal code does not punish Christian converts, yet as long as there is the article prohibiting Christianity in Kosatsu (tablet of laws put up in public places) Japan would look like a barbarous country not recognising freedom of worship, and therefore unworthy of being placed on the footing of equality with other nations. Hence, the said article should be struck out.' See Stead's *Japan by the Japanese*, p. 156.

(3) Already in 1872 there were six Japanese young ladies, brought by the Embassy, in American homes and schools, three of whom graduated from Vassar College, becoming in later years leaders of society in Tokio and wives of eminent men. In Japan, General Kuroda, Head of the Colonisation Department in Yezo had established a school for Japanese girls in Tokio, in which Miss Toewater (now Mrs Beukema of the Hague) and Miss de Ruyter of Brussels, ladies of the highest social rank and educational abilities, were teachers. In 1873, the first school for the education of the daughters of the gentry under the Department of Education was established in Tokio, under the charge of Mrs P. V. Veeder and Miss Margaret Clark Griffis. Out of this grew the Peeresses school. Besides a Women's University under President Naruse, and over two million girls in the 30,000 or so public schools, from kindergarten to the University, there are numerous private schools for girls. See index of Miss Bacon's *Japanese Girls and Women*.

(4) On the modern history of law in Japan see Georges Bousquet, *Le Japon de nos Jours*; Chamberlain's *Things Japanese*, with bibliography; the articles in T.A.S.J. and the German Asiatic Society of Japan, the pamphlets in English by the editor's former pupils in the Imperial University of Tokio, Dr Masujima and Professor Hozumi and Stead's *Japan by the Japanese*. The statistics of courts and prisons and civil and criminal law for 1903 may be found in the annual Resumé Statistique for 1906, pp. 118-138.

CHAPTER THIRTY-ONE

(1) See *Verbeck of Japan*, pp. 269-271.

(2) This was the time of the critical struggle between the old conservatives of feudal and medieval mind, and the progressives, and of the assassination of the latter by the former. See *The Mikado's Empire*, p. 574.

CHAPTER THIRTY-TWO

(1) Mr Brunton's plans were carried out in the years 1889-1896, when the harbour of Yokohama was enclosed by two breakwaters one and a quarter

miles long, and an iron pier nineteen hundred feet in length and connected with the railway to Tokio.

(2) Now Viscount Tadasa Hayashi. Born in Tokio in 1830, he learned English under Dr J. C. Hepburn. After nearly twenty-five years in Government service, mostly in the Foreign Office, he served as envoy in Peking and St Petersburg, and as Minister Plentipotentiary and Ambassador in London. He was especially active in securing the Anglo-Japanese alliance. He is now (1906) Minister of Foreign Affairs, an anti-militarist and member of the Hague Conference of 1899.

CHAPTER THIRTY-THREE

(1) Colonel C. O. Shepard, born in 1844, U.S. Consul at Yokohama, was for ten years U.S. Consul at Leeds or Bradford, England. Was knighted by the King of Portugal for his services in connection with the Maria Luz affair in Japan, by which the Chinese coolie traffic by Peruvians was put a stop to. (See *The Mikado's Empire*, p. 567). He was also honoured by the Chinese Government; author of *New York: Colony and State in the Revolution*, and numerous stories in the periodicals; was for four years Governor of the Soldiers' Home at Bath, N.Y.; for several years has been U.S. Government Special Inspector of Indian leases in the Indian Territory under the Department of the Interior.

CHAPTER THIRTY-FIVE

(1) The 'king' Sho-tai, deposed in 1879, was made a marquis in the new nobility of Japan and compelled to live in Tokio, while the Shuri castle was occupied by a company of Japanese soldiers. Sho-tai died in Tokio 19 August 1901. For descriptive and scholarly papers on the Riu Kiu Islands, see T.A.S.J. The population of Okinawa ken in 1904 was 469,203, or 233,361 males and 235,842 females. Japanese customs in dress, coiffure, manners and education as well as Christian ideas are increasingly prevalent.

CHAPTER THIRTY-SIX

(1) It is needless to say that the editor takes an entirely different view of the causes of Japan's renascence. See the chapter 'The Recent Revolutions of Japan', *The Mikado's Empire*, pp. 291-324; 'The Mind of Modern Japan' in the *Homiletic Review*, and Okakura's *The Awakening of Japan*, and Dr Nitobe's *Bushido*.

(2) One battalion of the 20th and the 10th Regiment, some Beloochis from Shanghai, and a detachment of Royal Marines were stationed at Yokohama, between 1863 and 1880, as many as 1,200 men being in camp at one time.

NOTES

BY HUGH CORTAZZI

Griffis' Introduction

(1) Adams, William (1564-1620) has been called the first Englishman in Japan. He was born in Gillingham, Kent, and came to Japan in 1600 as a pilot on the Dutch ship *Liefde* ('Charity') which foundered off the Usuki peninsula in Kyushu. Adams stayed in Japan and learnt Japanese. He acted as a builder of ships and as adviser/interpreter to Ieyasu, the first Tokugawa Shogun. He was granted samurai status, an estate near Yokosuka and had a Japanese wife. When the first British trading post was established at Hirado in Kyushu in 1613 he was employed as a go-between but was never trusted by the British merchants who thought he had gone 'native'. They referred to him as a 'Japanner'. For further information about Adams see, for instance, Richard Tames: *Servant of the Shogun*, Paul Norbury Publications, Tenterden, 1981. An excellent account of the British contacts with Japan in the seventeenth century is contained in Derek Massarella's *A World Elsewhere*. Yale University Press, 1990.

(2) Saris, John (1580-1643) was the commander of the *Clove* which came to Japan in 1613 to establish a British trading post.

(3) Cocks, Richard (1566-1624) was the head of the British trading post ('factory') at Hirado from 1613-1623.

(4) Mitford,. A. B., first Lord Redesdale (1837-1925) served on the staff of the British Legation in Japan from 1866-1870. See *Mitford's Japan* edited by Hugh Cortazzi, Athlone, 1985.

(5) Satow, Ernest (later Sir Ernest) (1843-1929) came to Japan in 1862 as a student interpreter and played a significant role as adviser to Sir Harry Parkes, the British Minister at the time of the Meiji Restoration. His fascinating account of his service in Japan during this period is contained in *A Diplomat in Japan*, first published in London in 1921.

(6) Aston, W. G. (1841-1911) studied Japanese as a member of the British 'Japan Consular Service' and became an outstanding Japanologist.

(7) Parkes, Sir Harry (1828-1885) was British Minister to Japan from 1865-1883. For details of his life in China and Japan see S. Lane-Poole and F. V. Dickens *The Life of Sir Harry Parkes*, 2 volumes, London 1894.

(8) Alcock, Sir Rutherford (1809-1897) was the first British Consul General and Minister to Japan, arriving there in 1859. His experiences in Japan are recorded in *The Capital of The Tycoon*, 2 vols, London 1863.

(9) Satsuma, a fief in Southern Kyushu, now Kagoshima prefecture.

(10) Mint. In 1868 the Japanese Government purchased the British mint in Hong Kong, but the machinery was destroyed by fire in 1869. New machinery was imported from Britain and Major William Kinder was appointed in 1870 as the first director of the mint.

(11) Railway. The first Japanese railway from Yokohama to Tokyo was completed in 1872. The chief engineer was Edmund Morrel, an Englishman, who died before the line was completed. For an account of the building of the railway see Grace Fox's *Britain and Japan, 1858-1883*. Oxford, Clarendon Press, 1969.

(12) Oyomei philosophy (*Yomeigaku* or the Wang Yangming School of Confucianism as expounded by Wang Yangming (1472-1529). emphasised that there was in all men 'good-knowing' i.e. intuition. It also stressed the union of thought and action.

(13) Okuma Shigenobu (1838-1922), a prominent Meiji politician who was twice Prime Minister of Japan and served in various cabinet posts including that of Foreign Minister. He founded Waseda University in Tokyo.

(14) Okubo Toshimichi (1830-1878), Satsuma samurai and leading figure in the Meiji Restoration.

(15) Ito Hirobumi (1841-1909) was one of the foremost leaders of Japan after the Restoration. He was the main author of the Meiji Constitution of 1889 and served as Prime Minister four times. He was created a Prince. He was assassinated at Harbin by a Korean nationalist.

(16) Inoue Kaoru (1836-1915), Meiji statesman who served as a leading Minister in various cabinets.

(17) Yokosuka is the main Japanese naval base west of Yokohama. The ship seems to have been either the *Aki* or the *Satsuma*.

(18) The Sino-Japanese war of 1894/1895.

(19) The Russo-Japanese war of 1904/5.

(20) *Yatoi* is used to mean an employee. Thus the *o-yatoi-gaijin* were the honourable foreign employees.

(21) 1897. The new treaties in fact came into force in 1899.

(22) Tycoonery. The word Tycoon comes from the Japanese *taikun*, literally great prince, and was another term for the Shogun.

(23) Harris, Townsend, (1804-1878), the first US Consul General in Japan. He arrived in Japan in 1856 and was stationed first at the small port of Shimoda on the Izu peninsula. He negotiated the commercial treaty of 1858 and was appointed Minister to Japan in 1859. He returned to the USA in 1862.

(24) Sanjo Sanetomi (1837-1891), Meiji statesman.

(25) Iwakura Tomomi (1825-1883), Court noble and Meiji statesman who led the famous Iwakura Mission to North America and Europe between 1872 and 1873. (The mission had left Japan on 23 December 1871 to cross the Pacific and arrived in San Francisco Bay on 15 January 1872.)

(26) Kido Koin (1833-1877), Meiji leader who played a prominent part in the overthrow of the Shogunate.

(27) Terashima Munenori (1832-1893), Satsuma samurai who was the Japanese Foreign Minister between 1873 and 1878.

(28) Sano Tsunetami, (1822-1902), founder of the Japan Red Cross who played a leading role in the introduction of Western science into Japan. He was also prominent in establishing the Japanese navy.

(29) Prince Arthur of Connaught was the son of the Duke of Connaught and grandson of Queen Victoria. He was sent to Japan in 1905 to present the Order of the Garter to the Emperor Meiji.

(30) MacDonald, Ranald (1824-1894), an American adventurer who jumped ship in 1848 and reached Japan where he was promptly arrested as foreigners were then prohibited from entering Japan. He claimed to have taught English to some Japanese while he was a captive. He was returned to the USA the following year.

Brunton Notes

CHAPTER ONE

(1) The supplementary convention was The Tariff Convention of 1866. Grace Fox (*Great Britain and Japan, 1858-1883*) page 370 commented as follows: 'British and French naval authorities recommended the sites most in need of lights and the apparatus required. Parkes sent these recommendations to the *Roju* (Elders) on 17 November 1866, urging the promotion of the work.'

(2) According to Brunton's account in his lecture to the Institution of Civil Engineers in 1876 entitled 'The Japan Lights', two of the three lights ordered from France were installed at Noshima and Kannonsaki under the supervision of French engineers stationed at Yokosuka.

(3) Trinity House is the British authority responsible for pilots, lighthouses and buoys.

CHAPTER TWO

(1) Vice-Admiral Sir Henry Keppel took command of the China Station in April 1867.

(2) Forty-two characters. Brunton means the *kana* syllabary of 48 syllables. *Katakana* rather than *hiragana* was used for telegrams.

(3) Grace Fox, *Britain and Japan 1858-1882* (page 376) noted that there was some opposition to the innovation of telegraph lines. '"Two-sworded roughs" hacked at some posts in the beginning and ignorant people often broke down the line so that guarding it was difficult. By Spring 1870, however, the single wire was hardly adequate to handle the demand for transmitting Japanese messages between Yedo and Yokohama.' By 1872 a line between Tokyo and Nagasaki had been completed and this 'made possible telegraphic communications with London through the Great Northern Telegraph.' Although Brunton was clearly involved from the beginning with this project the responsibility for its execution and for training Japanese staff fell to Mr Gilbert.

In a letter to the editor of *The Japan Weekly Mail* dated 9 March 1870 (printed in the issue for 12 March 1870) Brunton commented as follows on a letter in

L'Echo du Japon dated 5 March and signed Jules Furet: 'The Line of Telegraph is a measure carried out for convenience as well as for economy. The growing importance of Yokohama, and the centralisation of Government departments in Yedo caused an immense number of immediate messages to circulate between the two places. These having to be conveyed on horseback have been a growing inconvenience and expense which the telegraph was devised to remedy. The authorities had always in contemplation opening it to both Foreigners and Natives, and they have done so to the latter at, in my opinion, an absurdly low tariff. The result however is, and it is an interesting one, that messages are now pouring in, to the number of one hundred a day, at each end. Opening it to Foreigners is at present only delayed until interpreters can be found in whom sufficient confidence can be placed for the transmission of messages in English. The entire cost of the Telegraph including instruments, poles, cost of erection and a Telegraph office at each end has been $5,500, about one seventieth of the amount which M. Jules Furet gives for the enlightenment of his countrymen.'

CHAPTER THREE

(1) Brunton is referring to the Tariff Convention and a 'Convention of Improvement of Settlement, Racecourse, Cemetery, etc., of Yokohama', signed on 29 December 1866. This provided for replanning the settlement, paving the main streets, the drainage of streets and swamps and the construction of pavements (side-walks). Yokohama in the 1860s was like a Wild West frontier town (see Hugh Cortazzi's lecture to the Royal Society for Asian Affairs 'Yokohama Frontier Town 1859-1866' (*Asian Affairs* Vol. XVII Part 1 February 1986), chapter on Yokohama in Hugh Cortazzi's *Victorians in Japan* and Pat Barr's *Coming of the Barbarians*).

(2) Anna D'A in *A Lady's Visit to Manilla and Japan*, London 1863, page 253 commented that 'The houses...are very slightly built, mostly of wood, many of the poorer ones being composed only of a light bamboo framework, covered with thick mud....'

(3) The Bluff (Yamate) in due course became the most favoured place for foreigners to reside.

(4) Shimoda, small port towards the end of the Izu peninsula, to the west of Yokohama. In those days it was a remote spot and difficult to reach by land.

(5) There was no vehicular traffic on the main roads in Japan until after the treaties were concluded in 1858. All traffic on the main Tokaido route from Edo (Tokyo) to Kyoto was either on foot, or on horseback or in a palanquin.

(6) *The Japan Weekly Mail* for 2 April 1870 reported: 'Mr Brunton is rapidly becoming the presiding genius of the settlement. He has transformed the bund from a Slough of Despond into a fine road, he is draining the streets and laying them with metal, which, if not the best in the world, is the best that can be procured, he has sketched out an excellent scheme for a water supply, and finally has promised his assistance, if the Government will allow him time to expend his irrepressible energies on it, towards the lighting of the settlement.' Associated with Brunton in this latter scheme was Signor Beato, the photographer.

The Japan Times Overland Mail for 30 December 1869 contained a long unsigned article entitled 'The Drainage'. The writer asserts that the residents of Yokohama would do well to assist 'the execution of Mr Brunton's scheme'.

It then rebuts a criticism of his scheme in the columns of the *Herald* which, it claimed, misrepresented his plans. 'It is not true that in his proposed drains "one foot in 400 is by no means uncommon", and it is quite a mistake to say that the sewage will make on the shore of the Bund "a pestiferous ooze at dead low water, which washing up and down with the ebb and flow of each tide will become a floating nuisance". There is only one main drain which has a fall of one foot in 400 and this, by a little manoeuvring of the levels, can be increased to one in 300: few of the other drains are less than one in 200 and some have as great a fall as one in 50.... The "pestiferous ooze" is purely imaginary, nothing but liquid matter can in any case pass the trap wells sunk at intervals along the bund, gratings at these points preventing any solid matter, even sticks and straw, from passing into the outflow pipes. The mouths of these pipes will open half way between high and low water marks, the rise and fall is 5¾ feet in ordinary spring tides and 3 feet in ordinary neap tides, the pipes will thus be flushed for hundreds of feet every tide and there cannot be any collection of offensive matter along the bund whatever.'

The article then goes on to discuss whether residents will have their house sewage carried away by the drains. This was a matter for each householder but they would have to 'do so in the way pointed out, by the apparatus described by Mr Brunton. We are of the opinion that until a proper water supply is provided...it will be better to use other means for this. The recently patented invention of earth boxes is in use in a few houses and is found to answer admirably and we see no reason why, to supplement what it is now doing, the Japanese Government should not organise a proper and effective *corps* of scavengers with the iron carts and lime deoderiser in use in Europe, by means of which our outhouses might be effectually cleaned nightly. This method has the great advantage of cheapness....'

The article noted that Brunton's scheme was confined to the foreign settlement only. 'However well drained our own settlement may be, danger of course must always exist from our proximity to an offensive and dangerous neighbourhood.' The article concluded: 'the Engineer will welcome suggestions for their [i.e. the drains'] reasonable extension and improvement: but it must be remembered by those who have suggestions to make that he *is* an Engineer and, consequently, a man of hard facts and figures, and that he cannot be expected to entertain any but proposals which are practical, practicable and sound.'

The article, if not written by Brunton, clearly reflected his views.

On 27 January 1871 Brunton wrote to *The Japan Weekly Mail* (edition of 28 January 1871) complaining that lithograph plans of the settlement were on sale at the Municipal office. These were 'very fairly executed' copies of the plan prepared in his office. Publication had been made 'virtually without my knowledge and without its authorship being in any way recognised on it.'

Nevertheless, Brunton continued to help with the development of the settlement as is clear from an article in *The Japan Weekly Mail* dated 18 May 1872. The article noted that 'The levelling of the swamp is nearly finished; the cricket ground is in the course of construction, and the time is fast approaching when the Government will sell large portions of the unenclosed land for building purposes. Yokohama, already cramped and confined on all sides, will then be worse off than ever, and those who live in the settlement will be debarred from any recreations such as is within reach of the residents on the bluff.'

It went on to point out that under the Convention of 1866 the Japanese Government had agreed 'to enlarge, lay out and plant as a public garden, to

176

be used by both foreigners and Japanese, the site of the old Kosaki Machi....'
This gave a good claim upcn the Japanese for the construction of gardens on
the swamp. The garden 'would be about 20,000 tsubo or 16²/₃ acres in extent,
the cricket ground in the centre taking up a space of about ninety square
yards, thus leaving ample room for laying out paths and flower beds. We
believe that Mr Brunton has already made a plan for laying out the gardens
which has been accepted.' The problem lay in the allocation of costs.

CHAPTER FOUR

(1) Brunton's 'Scheme For A Water Supply To Yokohama' was published in
The Japan Weekly Mail for 12 March 1870. *See Appendix I.*
Brunton's plans aroused a number of comments in the local English language
press.
The Japan Times Overland Mail for 30 December 1869 drew attention to the
health dangers from sewage matter: 'Living as we do, in close contact with
so large and daily increasing a native population, and one, too, remarkable
for two antipodal qualities, a love of cleanliness inside their houses, conjoined
with an almost total disregard of the commonest sanitary precautions outside,
it is a matter for wonder that, up to the present time, so little should have
been attempted towards securing the benefits which a constant water supply
could not fail to bring in its train....' Brunton had pointed out that without
a water supply the drainage works could not be expected 'to carry off that
portion of sewage matter which is most offensive, and that the dissolution
and removal of which is therefore most necessary.'
An article in *The Japan Weekly Mail* for 19 March 1870 commented that
Brunton's water supply scheme would cost some $220,000 (adding an
additional $10,000 contingency). At 10% interest, this would amount to
$22,000 per annum. 'For this sum we should have a supply of 390,000 gallons
per day for the native town and the foreign settlement, the latter receiving
120,000 gallons by a reduced piping after the former was supplied. This supply
would be laid on on the *constant system*, with a pressure due to the elevation
of the fountain head which would raise the water to the tops of the highest
houses, and admit of a jet being thrown over them in cases of fire. Water
columns to the number of one hundred and fifty would be placed at convenient
distances in the streets, and hydrants at the distance of one hundred yards
from one another.'
The article then discussed the advantages this would bring in terms of fire
insurance costs. To these the author added: 'Of the saving which would take
place in our own houses for coolie hire, ropes, buckets and all the etcetera
which are the sources of irritating charges, and often more irritating extortions,
we have said nothing, because they will occur to every house-holder and
although important, are minor considerations.' The article called on 'a few
energetic members of the community' to get together with influential Japanese
to ensure that the necessary funds were found to get the work started. It ended
with this encomium on Brunton's plans: 'We should be doing even less justice
to Mr Brunton's scheme than we have done, if we failed to notice the care,
the modesty, good sense and conscientiousness which pervades the matter in
which it has been drawn up.'
The Japan Weekly Mail for 19 November 1870 reverted to Brunton's scheme
and was critical of the community's failure to progress the scheme. It noted:
'The water at present used in Yokohama is of a most deleterious character.
It is drawn from wells which depend for their supplies upon a species of

surface drainage....' Drawing a parallel with London it asserted that epidemics were inevitable. 'With a native population, whose habits of cleanliness are none of the best, the wells which draw their supplies from surface drainage - and nearly all do it - almost of necessity must contain in solution large quantities of organic matter, of such ʋ nature that it is a wonder - almost a miracle - that Yokohama has not been visited with an epidemic long ere this.'

The article then reverted to Brunton's calculations noting that there were in 'March last about 600 Europeans, 1,200 Chinese and 18,889 natives, but allowing for an increase in population he estimated the total at 23,237 persons.' Brunton reckoned that the costs could be met by charging native houses $4, Chinese $6 and European households $7. The article noted one weak point in Brunton's scheme namely the need for 'some means for preventing the unlimited use of the water which would certainly take place among the Japanese.' The Japanese Government should assist by 'compelling the natives to pay their quota and to use the water supplied to them by the mains.' The article concluded: 'As it is now we are compelled every day to run the risk of severe illness; we drink in poison without suspecting its virulence or its potency, and we daily absorb into our bodies such elements as must affect our bodily health. This state of things should not be allowed to exist for one single moment, and we earnestly entreat the community to unite in one effort to procure an abundant and constant supply of water.'

(2) An article signed R.H.B. on the subject of 'Lighting' appeared in *The Japan Weekly Mail* for 20 September 1873. The text of this article which is relevant to the question of providing lights for Yokohama is at Appendix II.

CHAPTER FIVE

(1) The iron bridge whose construction in Yokohama was planned and executed by Brunton was the second, not the first, iron bridge in Japan. The first iron bridge had been built in Nagasaki in April 1868. It had been constructed by one Motoki Akira who had studied Dutch engineering. He had been adopted into one of the families which provided Dutch interpreters for the Dutch in Dejima.

In his letter to *The Japan Weekly Mail* dated 9 March 1870 (in the issue for 12 March 1870) Brunton replied to criticisms in *L'Echo du Japon*, signed Jules Furet. He wrote: 'He [Furet] asks where the necessity lay for replacing a wooden bridge by an iron one at an absurd expense to a Government crippled by debt?... The iron bridge at Yoshida was erected at the request of the Japanese authorities purely on economical grounds, the wooden one which it replaced being so frail that serious thoughts were entertained of closing it before we could get the iron bridge completed. The immense cost of renewing and maintaining their wooden bridges was the inducement which led the authorities to attempt this iron one, and though there were serious difficulties from the beginning of its construction to its finish, its cost has not been extravagant, amounting to about $15,000, and the manner in which it stood the tests to which it was subjected shewed it to be quite up to its calculated strength.'

CHAPTER SIX

(1) There are now two *fu* namely Kyoto and Osaka. Tokyo which was a *fu* is now a *to* meaning capital city. Japan is described as having 47 prefectures consisting of one *to* (Tokyo), one *do* (Hokkaido), two *fu* (Kyoto and Osaka), and 43 *ken* (ordinary prefectures).

178

(2) Under Article III of the Anglo-Japanese Treaty concluded by Lord Elgin in 1858 Osaka had been due to be opened on 1 January 1863 for residence by British subjects, although they were not granted the same rights of leasing land, purchasing buildings and erecting dwellings, which they were granted in the neighbouring treaty port of Kobe (Hyogo). In fact the opening of Osaka was postponed until 1 September 1868.

(3) Godai Tomoatsu (1825-1885), a Satsuma samurai who went to Europe after the British bombardment of Kagoshima in 1863, later became prominent in business. In 1878 he took the lead in establishing the Osaka Chamber of Commerce.

(4) The railway between Kyoto and Kobe via Osaka was completed in 1876. The Kobe-Osaka stretch was completed in 1873. Osaka was overshadowed as a port by Kobe until after the Second World War.

CHAPTER SEVEN

(1) Niigata should have opened as a treaty port from 1 January 1860. In fact it was not opened until 1869. The first British Consul, James Troup, was appointed in August of that year. British Consuls continued to reside there until 1877, but the port was never active in foreign trade during the nineteenth century. In 1877 there were only six British residents in Niigata.

(2) Okodzu would now be romanised as Okozu. Taradomari is a misprint for Teradomari.

(3) Dutch engineers had considerable practical experience in this sort of work.

(4) Yamao Yōzō.

CHAPTER EIGHT

(1) Sado, part of Niigata prefecture, is the largest island in the Sea of Japan with a circumference of 217 kilometers. From the thirteenth to the sixteenth century the island was used as a place to which prominent people were banished. These included the ex-Emperor Juntoku and the Buddhist priest Nichiren. Gold and silver mines were opened on the island in 1601.

CHAPTER NINE

(1) The source of this quotation has not been traced. Perhaps Brunton was quoting one of his own comments.

(2) Horatio Nelson Lay. See Grace Fox's *Britain and Japan, 1858-1883*, pages 386-393, Hugh Cortazzi's *Victorians in Japan*, pages 317-336 and Neil Pedlar's *The Imported Pioneers*, pages 95-99.

(3) See E. G. Holtham's *Eight Years in Japan, 1873-1881*, London 1883.

CHAPTER TEN

(1) For a survey of Japanese map-making and interchange with the West see Hugh Cortazzi's *Isles of Gold; Antique Maps of Japan*, Tokyo 1983.

(2) Sidotti, Giovanni Batista (1667-1714) was the last Roman Catholic missionary to reach Japan after the Tokugawa ban on Christianity. He landed in Kyushu in 1708 and was immediately captured. He was imprisoned in the so-called *Kirishitan Yashiki*. He managed to convert two of his caretakers and was then confined in a hole in the ground in which he died. He was questioned at length by the famous Confucian scholar Arai Hakuseki (1657-1725).

(3) Ino Chukei (Tadataka) (1745-1818) was the first surveyor to use Western scientific methods.

(4) The generally accepted system of romanisation is the Hepburn system promoted by James Curtis Hepburn (1815-1911), an American medical missionary in Japan.

(5) Isabella Bird's *Unbeaten tracks in Japan* was first published in 1880. It is a classic piece of travel literature and describes a journey which took Miss Bird with her Japanese interpreter through central Honshu to Niigata and then north to Hokkaido.

(6) Brown, Albert Richard. See Lewis Bush's *The Life and Times of The Illustrious Captain Brown: A Chronicle of the Sea and of Japan's Emergence as a World Power*, published in Tokyo and Rutland, Vermont, in 1969.

(7) Brunton's practical approach to training led him into a controversy with Henry Dyer, the Scottish principal of the Engineering College (Kobu Gakko). *The Japan Weekly Mail* for 18 September 1875 contained a long review by Brunton of a book entitled *Indian Public Works and cognate Indian topics* by William Thomas Thornton, Secretary for Public Works in the India Office in London, published in London in 1875. In the last part of this review Brunton turned his attention to Japan. He first criticised the Engineering College in the following terms:
'With such opinions as these before us it is difficult to resist the conclusion that too much importance is given, in the Engineering College in Yedo, to theory. What we imagine the Japanese desire, and what will probably be of the greatest value to them is to learn how to build a bridge, how to form a railway, or how to make a harbour, and they in all likelihood desire to gain this knowledge in the shortest way possible. The "valuable accomplishment" of knowledge in the higher branches of Physics and Mathematics is not likely to be of much service to them for many years to come....
'The first and most important object for the Government is naturally to obtain Engineers in whom they can place reliance. Having procured these, the fewer restraints placed upon their actions in the professional discharge of their duties the better....' These thoughts led him on to vent his frustrations against the Government and the bureaucracy under which he had to operate:
'In this country, Engineers have but little communication with the Government, their actions are controlled and closely watched by subordinate officials, the carrying out of the smallest details in an authorised work is criticised and interfered with; all suggestions have to be laid before them; the decision rests as to whether these are worthy of being referred to the Government or not.... On the supposition that an Engineer requires to be so closely watched who, in the conduct of Public Works in this country are the watchers? Are they the superiors of the Engineers in capacity or integrity?.... We question very much if any intelligent or conscientious statesman in the Government will claim superiority over, or even equality with, the Engineers, for the officials under whom they are placed. And we would desire strongly

to urge the abolition of a system which is incongruous, paradoxical, and destructive of anything approaching an efficient prosecution of Public Works in Japan.

'Public Works, we are constrained to say, do not hold their proper place in this country. On any shortness of money in the National Treasury, the first attack is always made on the expenditure incurred for these. So much so, that, latterly, their prosecution has resembled the progress of a hare, or the breaking of rollers on the sea coast; at one moment, rushing, with impetuosity along the strand, to be followed by serene quietude, so soon as their energy is expended. We doubt exceedingly if the various other methods which the Government has adopted for spending its money will have so great a tendency to increase the material prosperity of the country, or her influence abroad, as the execution of these internal improvements which constitute one of the chief requirements of civilisation.'

Brunton's criticism of the Engineering College provoked an outburst from Henry Dyer, the Principal. This led to a riposte from Brunton in a letter dated 26 September 1875 which appeared in *The Japan Weekly Mail's* edition of 2 October 1875. [See Appendix 3 for the full text of the exchange.]

CHAPTER ELEVEN

(1) Kinder, Major William, director of the Japanese mint 1870-1875. See Grace Fox's *Britain and Japan, 1858-1883*, pages 402-406.

(2) After the opening of the treaty ports the first currency used for trade was the Mexican silver dollar. In December 1870 it was announced that a one yen silver coin would be issued with 416 grains of silver equivalent to the silver content of the Mexican dollar.

(3) Sanjo Sanetomi (1837-1891).

(4) Date Muneki (1817-1882) played a conspicuous role at the time of the Meiji Restoration.

(5) Okuma Shigenobu (1838-1922).

(6) The Sawa and Kujo (not Kajo) were *kuge* (princely) families from Kyoto.

(7) Howell, W. G. see Grace Fox's *Britain and Japan, 1858-1883*, pages 429-430 and 449-450.

(8) It seems unlikely that this was the first salute fired from Japanese shores.

CHAPTER TWELVE

(1) Brunton delivered two lectures to the Asiatic Society of Japan in Yokohama on 'The Constructive Art of Japan'. These lectures were given on 22 December 1873 and 13 January 1875. See Appendices 5 and 6.

CHAPTER THIRTEEN

(1) *The Far East* (Volume 1 No. VI dated Thursday 16 August 1870), a fortnightly illustrated magazine, recorded an explosion on the 'City of Yedo', a steamer which plied between Yokohama and Tokyo before the railway was completed. The accident took place on 1 August 1870. The steamer was built in Yokohama. The magazine noted:

'The boiler was of high-pressure, and of peculiar construction, and had been recently overhauled, and had new pipes throughout, by professional boiler-makers. For many months she had been running with somewhat defective tubes, but with care, the boiler had held out, and we doubt not that with care it would have continued to do its work without danger; but the carelessness or ignorance of a man, evidently no engineer, but who was permitted to act as such, led to a disaster of a most fatal character. On Monday afternoon, the 1st of August, just as the vessel was unmooring at the *Hatoba* [pier], Yedo, to start for Yokohama, the boiler blew up, hurling nearly or quite 150 persons into the water, wounding almost every one, and killing an American missionary Mr Cornes, his wife, one of their infant children and their English nurse-maid, Mr. Cassidy the acting-engineer, and very many Japanese.... The boiler had exploded laterally, parting in the middle. The forward part dashed through everything into the fore cabin, killing all therein, with the exception of a little baby who was in it nurse's arms - the nurse was killed. The other part of the boiler blew out astern, tearing up the deck and everything else, and was so completely broken into little bits, that no bulky portion of it could be found....

'The steamer lies on her starboard side with her bow up-stream - her funnel and all connected with it, the bridge, and a great portion of her decks blown out of her, and the paddle boxes broken.'

CHAPTER FOURTEEN

(1) *H.M.S. Manila*, according to Grace Fox, not *Manilla*.

(2) Mizuno, Izumi no Kami i.e. 'governor' of Izumi, one of the five provinces in the so-called Kinai around Kyoto. This was a title or rank and did not imply that Mizuno actually had a territorial role.

(3) Mikomotoshima.

(4) Oshima (Vries Island) is a volcanic island to the east of the Izu peninsula. It is administered as part of Tokyo prefecture. It has a mild climate and many hot springs.

CHAPTER FIFTEEN

(1) Nagasaki was the port where the Dutch merchants were confined on the artificial island of Dejima during the Tokugawa period. It was one of the first treaty ports.

(2) Inoue Kaoru (1836-1915), famous Meiji leader, came from Choshu, now part of Yamaguchi prefecture in western Honshu.

(3) Glover, Thomas Blake (1838-1911), founder of Glover and Company, was particularly active in the last years of the Shogunate in procuring ships and equipment for the Satsuma and Choshu fiefs. He is believed to have assisted Ito Hirobumi, Inoue Kaoru and Godai Tomoatsu to go abroad secretly when travel outside Japan was forbidden by the shogunate.

(4) Takashima coal mine was discovered in the early eighteenth century. The Saga fief in 1866 cooperated with Glover and Company in the management of the mine.

CHAPTER SIXTEEN

(1) Brown. See Chapter X note 6.

(2) The text of this paper was reproduced as an appendix to Lewis Bush's *The Life and Times of the Illustrious Captain Brown*. [See Appendix 7.]

CHAPTER SEVENTEEN

(1) This is a reference to the Namamugi incident (see Introduction).

(2) Shimazu Saburo or Shimazu Hisamitsu (1817-1887) was the *de facto* ruler of the Satsuma domain in the years leading up to the Meiji Restoration. The nominal head of the fief was his nephew, Shimazu Tadayoshi, son of Shimazu Nariakira.

(3) According to Grace Fox in *Britain and Japan, 1858-1883*, page 337, a spinning and weaving factory, purchased from Platt and Company of Manchester, was established in about 1866. It employed more than 200 workmen.

(4) It seems surprising that the Japanese should have imported Staffordshire pottery considering the development of their own ceramic industry, but there had hitherto been little or no demand in Japan for dishes suited to Western cuisine.

(5) The meal given to Sir Harry Parkes and his party in 1866 was endless with over 40 Japanese courses. See Hugh Cortazzi's *Victorians in Japan*, pages 254-255.

(6) See Hugh Cortazzi's *Dr Willis in Japan*.

CHAPTER EIGHTEEN

(1) Oneida. See 'The Loss of the U.S.S. 'Oneida' in Harold S. Williams' *Shades of the Past: Indiscreet Tales of Japan*, Tokyo and Rutland, Vermont, 1958.

CHAPTER NINETEEN

(1) The Empress Eugenie was the consort of Napoleon III of France.

CHAPTER TWENTY

(1) Iwakura Tomomi (1825-1883), leader of the Iwakura mission (see Chapter XXVII).

CHAPTER TWENTY-ONE

(1) The Dutch name of the ship was the Liefde ('Charity').

CHAPTER TWENTY-TWO

(1) See Grace Fox's *Great Britain and Japan, 1858-1883*, pages 280-283.

CHAPTER TWENTY-THREE

(1) Sano Tsunetami (1822-1902) see note 28 to Griffis' Introduction.

(2) Satsuma rebellion. See for instance A. H. Mounsey's *The Satsuma Rebellion: An Episode of Modern Japanese History*, London, 1879.

CHAPTER TWENTY-SIX

(1) The Dajokan (the Great Council of State) was originally established in the late seventh and early eighth centuries. In the early Meiji period it became the main organ of government with the Dajodaijin acting as a sort of Prime Minister.

(2) The Duke of Edinburgh was not the first European the Emperor had met. In March 1868 the Ministers of the main powers, including Sir Harry Parkes, the British Minister, had been received by him in audience in Kyoto. Sir Harry Parkes had also presented his credentials to the Emperor in Osaka in May 1868. The Duke of Edinburgh was, however, the first royal prince from Europe to be received by the Emperor. For an account of this visit see Hugh Cortazzi's *Mitford's Japan*.

(3) *The Japan Weekly Mail* for 23 December 1871 carried the texts of the Emperor's remarks and of Brunton's reply.

CHAPTER TWENTY-SEVEN

(1) For a summary account of the Iwakura Mission and a select bibliography on the mission see the entry in Kodansha's *Encyclopaedia of Japan*, Tokyo 1983. See also 'The Itinerary of the Iwakura Mission to Britain' by D.W. Anthony and G.W. Healey, University of Sheffield Occasional Papers, No 1, 1987.

(2) Among the female members of the Mission was Tsuda Umeko (1865-1929), who later founded Tsuda College for women (now Tsuda University) in Tokyo.

CHAPTER TWENTY-EIGHT

(1) According to *The Japan Weekly Mail* for 27 April 1872 Mr and Mrs Brunton and two children left for Hong Kong on the *S.S. Phase*, a French steamer.

(2) See Marie Conte Helm's *Japan and the North East of England from 1862 to the present day*, London 1989.

CHAPTER TWENTY-NINE

(1) Totomi province is part of what is now Shizuoka prefecture (around Hamamatsu). Kodansha's *Encyclopaedia of Japan* (1983) in an article on 'Petroleum' says: 'The production of crude oil in Japan has been on record since 1875'. In the first half of the twentieth century exploration was conducted in the prefectures of Niigata, Yamagata and Akita. W. E. Griffis in his *The Mikado's Empire* noted that 'In 1874, 107,243 gallons of excellent petroleum were produced. With American methods of drilling, pumping and refinery, the yield and area of trial are increasing', page 604 of the edition published by Harper and Brothers, New York in 1876.

CHAPTER THIRTY

(1) The *Onna Daigaku* (The Greater Learning of Women) was published in 1716. It has been widely attributed to the Confucian philosopher Kaibara Ekken (1630–1714).

(2) In 1862 some 60,000 'hidden Christians' (*Kakure Kirishitan*) who had managed to retain their faith, in part at least, despite some 250 years of persecution, were discovered in the Nagasaki region of Kyushu. The Anti-Christian edicts were still in force and the *Kakure Kirishitan* were either jailed or exiled. In 1873 as a result of pressure from abroad religious sanctions were withdrawn and the exiled peasants were allowed to return home. Freedom of religion in Japan was not specifically granted until the promulgation of the Meiji Constitution in 1889.

(3) According to *The Japan Weekly Mail* for 5 April 1873, Mr and Mrs Brunton, Mrs Miller and two children, Miss Satchell and Miss Smith arrived on 5 April 1873 in Yokohama on board the French steamer *Menzaleh*. Presumably the two children belonged to the Bruntons, Mrs M was Mrs Miller and she had been helping to look after the children on board the ship.

(4) Boissonade de Fontarabie, Gustave Emile (1829–1910), a French legal scholar, went to Japan in 1873 and stayed until 1895. He contributed to the compilation of the Japanese penal and civil codes.

CHAPTER THIRTY-ONE

(1) Brunton did not fully appreciate the economic difficulties facing the Meiji Government. These are described in Chapter XV 'Finance and Society' in W. G. Beasley's *The Meiji Restoration*, Stanford 1972.

CHAPTER THIRTY-TWO

(1) *The Japan Times Overland Mail* dated 30 December 1869 carried an article entitled 'The English Hatoba' [pier]. This began with these critical comments:
    'If anyone wishes to see obstruction to trade reduced to scientific precision and magnified by insufferable perversity, we recommend them to visit the Custom-house wharf at the English Hatoba, and we promise that nothing shall be wanting to their complete gratification. We can guarantee that within the space allotted to the grim satisfaction they seek, more means can be found to produce loss, inconvenience, delay and vexation than within a similar space in any commercial port in the world....
    'In the first place, the sea-wall of the wharf is nearly perpendicular, and boats which have to be discharged there must either take their turn at the solitary flight of steps leading down to the water, which are at present blocked up with stones, or discharge their cargoes by means of planks, reaching at a perilous angle from the wharf to the lighter, which they are graciously permitted to use, but which are not furnished by the Custom-house. In order to facilitate the landing of the blocks of stone which the Government requires for the erection of new godowns on the wharf, but which have no earthly business to be brought there at all, a rude platform or jetty has been thrown out, and though urgently required for the landing of foreign goods, it is, with singularly indelicate irony, monopolised for a purpose wholly alien to the objects of the wharf, and productive of vexations delays in its legitimate business....'

(2) *The Japan Weekly Mail* for 13 February 1875 contained a letter signed with the initial H (Brunton's second initial) 'On the subject of the Harbour Scheme'. This read in part:

'There can be no doubt that the construction of such a work would facilitate the operation of the discharge and loading of vessels here, but from what has appeared as the experience derived from other works of similar character elsewhere, I doubt whether it would prove remunerative to those who carry it out; if constructed by Government (and it may [be] presumed no unaided private enterprise will undertake the matter,) it would be done more with a political view of assisting the commerce of the country in a public way than from a speculative spirit of making it pay.'

(3) *The Japan Weekly Mail* for 21 March 1874 carried an account of the Emperor's visit. According to this account 'The Imperial Party started from Yedo by the 3 o'clock train and arrived at 4 o'clock at the Yokohama station whence they at once proceeded to the Lighthouse establishment....

'The dress of the Empress and her ladies was gorgeous and picturesque, though it was marked by combinations of colours which our own views do not sanction. His Majesty arrived on horseback attended by a body-guard, while the Empress and her suite followed in carriages....

'The Imperial party were ushered into the reception chamber – a room used as a school-room attached to the establishment, and the Emperor took his place at a small table, to the left of which, and separated from it by a few feet, was another, similarly covered with rich native drapery, at which the Empress took her place....'

After viewing the oil refinery the Emperor's attention was attracted to 'a small experimental retort'. 'The Iron, Oil and General storehouses were then visited, and seemed to excite approbation for the excellent order in which they appeared.' In the machine shop the party saw 'a vertical saw capable of cutting a log 2 feet square into 24 planks at one operation'. They also viewed a 'general joiner', a couple of lathes and a hydraulic press used to test the quality of bricks and cement.

The party visited the 'Experimental Lighthouse' whose three floors were used for training purposes. The report noted that 'The lantern which has been erected on the Experimental Lighthouse was built entirely in Yokohama'. This apparatus was for use at Inubogasaki (at the northern point of Chiba prefecture). It was 'what is technically known as a 1st order, half-minute, revolving, dioptric apparatus. It is of the largest size made, being about six feet in diameter and nine feet high. It has eight faces or sides, each of which gives out a beam of parallel rays of light. It makes a complete revolution once in four minutes, so that each face, and the beams emitted by it, come before the eye of the observer every half minute. It is dioptric only in so far as it is entirely made of glass; but the central discs only of the apparatus are purely dioptric – that is to say, that they parallelise the rays by refraction alone; while the prisms above and below the disc are catadioptric, or partially refract, and partially reflect, the rays'. The lamp used was one of Doty's 4-wick burners.

The Emperor examined the various lamps 'for some time' in the darkened lower floor of the lighthouse. He then 'ascended the stairs and entered the inside of the apparatus, where the effect is very dazzling and peculiar'. His Majesty also examined an iron lighthouse in the course of construction and in the drawing offices asked various questions about the working drawings which were shown to him.

'After taking dinner, His Majesty, the Empress, and the party enjoyed the scene presented from the balcony of the house where they were entertained. The district in which it lies was bathed in the flood of light thrown by the great lamp which His Majesty had just inspected.'

On the following day their Majesties visited the Gas Works before returning by train to Tokyo.

(4) Hayashi Tadasu (1850-1913) had accompanied the Iwakura Mission.

CHAPTER THIRTY-THREE

(1) Aston, W. G. (1841-1911) was a member of the Japan Consular Service.

(2) St John, Captain H. C., R.N. was the author of *Notes and Sketches from the Wild Coasts of Nipon* published in Edinburgh in 1880.

(3) *The Japan Weekly Mail* for 18 October 1873 noted that the Japanese Government had 'been distinguished in the prize list of the Vienna Exhibition by "Honourable Mention" (a Diploma) for the lighting of their coasts, a fact creditable and gratifying both to the Government and the Department over which Mr Brunton presides'.

CHAPTER THIRTY-FOUR

(1) *The Japan Weekly Mail* for 11 March 1876 in reporting his departure described him as: '...a zealous, intelligent, sturdy and honest servant of this Government, whose works will remain for generations as beneficent monuments of his labours - pillars of cloud by day and fire by night. Mr Brunton's services were acknowledged just prior to his departure by a complimentary letter from the Minister of Public Works and a present of two thousand yen from the public purse. The Japanese act handsomely in these matters.'

Brunton left Japan on the steamship *Oceanic* for San Francisco on 10 March 1876. Mrs Brunton had left separately on the British steamer *Bombay* for Hong Kong on 15 December 1874.

CHAPTER THIRTY-FIVE

(1) Brunton gave a fuller account of his trip in a lecture to the Asiatic Society of Japan on 19 January 1876. This is given in Appendix 4.

(2) Naha.

(3) Fu-chou.

(4) *Alceste*, H.M.S. See John M'Leod's *Voyage of His Majesty's Ship* Alceste *along the Coast of Corea to the Island of Lewchew with an account of her subsequent shipwreck*, London 1818. Also Hall, Captain Basil's *Voyage to Loo-Choo and other places in the Eastern Seas in the year 1816*, Edinburgh 1826.

(5) Chamberlain, Professor Basil Hall, published in 1895 an 'Essay in aid of a Grammar and Dictionary of the Luchuan Language'.

CHAPTER THIRTY-SIX

(1) Brunton is presumably referring to Gilbert and Sullivan's *The Mikado*.

First *published in* THE JAPAN WEEKLY MAIL, 12 March 1870.

SCHEME FOR A WATER SUPPLY TO YOKOHAMA

[NOTE: The original spelling of place names has been retained throughout.]

PRESENT SUPPLY

THE NECESSITY for a water supply other than that now existing, is severely felt in the Native portion of the settlement, and an agitation has been going on for some time to improve it.

The impurity of the water at present used is a matter which must sooner or later attract the attention of the residents in Yokohama, both Foreign and Native.

The wells from which it is procured range from 12 to 20 feet deep, and are sunk in close proximity to open drains, stables, privies and ash pits, and as this supply comes principally from their immediate neighbourhood, they being so shallow that their drainage area must be of very limited extent, there can be no doubt that along with the water percolating into them through the veins of the earth, a large quantity of the obnoxious matter which exists around them must also find its way.

As Yokohama grows older the filth will accumulate in the wells, the passages through which they are supplied will themselves become contaminated and the results will speedily become disastrous.

The want of water in cases of fire is also an important consideration which requires remedying, the wells being only like small cisterns which may be emptied by a pump in a few minutes, taking hours to refill.

Further, the supply being dependent on the rainfall, from so confined an area in dry weather, it is very liable to fall, and in seasons of long and excessive drought, Yokohama might become entirely waterless.

PROPOSED NEW SUPPLY

The general principles on which the scheme is grounded are these. To supply sufficient water for the domestic and public wants of all the native town, including the adjoining districts of Yoshida, Nunge, Homura, and Nokamura, as well as the Foreign part of the settlement on the *constant system* - that is to have the Pipes constantly full, leaving people to use as much water as they choose, and so obviate the necessity for tanks in the dwellings - turn keys, &c. To bring the water at high pressure from a reservoir containing a few days supply, which will be sufficiently high to enable the water to rise by its own gravitation to the tops of the highest houses, and also to throw jets over them in cases of fire. To have the water thoroughly purified by an approved process of filtration, before admission to the reservoir.

Quantity - By information received from the Municipal Director, the population to be supplied is at present as follows:-

```
Natives, Yokohama  ............................................  12,960
   ,,     Yoshida, Nunge, Homura and Nokamura  .....   5,929
                                                        ──────
             Total Natives  ...........................  18,889
Chinese  ...........................................................   1,200
Europeans  .......................................................     600
                                                        ──────
             Total  .......................................  20,689
                                                        ══════
```

It will be necessary, however, to make allowance for a probable increase in these numbers, say:-

```
Natives   10 per cent  .............................................  20,777
Chinese   30    ,,     .............................................   1,560
Europeans 50    ,,     .............................................     900
             Total  .......................................  23,237
                                                        ══════
```

London is supplied with 30 Gallons per inhabitant per day, Liverpool with 24 Gallons, Glasgow new waterworks are capable of supplying 50 Gallons per inhabitant per day, and Manchester 42 Gallons.

In all these towns, more especially in the two latter, there are manufacteries and other works using large quantities of water, which must be considered in judging of the amount necessary for the wants of Yokohama. The habits of the people are also matters for consideration. The above quantities of course include all that is used for water closets, baths, or washing purposes. From our supply to the native population, may safely be excluded the quantity required for water closets, and the washing and bath water will also be less than for European requirements, ten gallons per inhabitant per day being a quantity which is seldom likely to be exceeded. The European supply on the other hand should have a liberal allowance for baths and washing, and may be put down at 40 Gallons per inhabitant per day, being 10 Gallons more than is supplied to London.

The quantity will at these figures be as follows:-

```
Natives    20,777, at 10 Gallons  ............................  207,770
Chinese     1,560, at 20     ,,        ...................   31,200
Europeans     900, at 40     ,,        ...................   36,000
                                                        ──────
                                                          274,970
```

Say, 300,000 Gallons per day.

Height of Fountain Head - The height above the highest points of the settlement should be as follows:-

Houses - 30 feet.

Loss of head from friction and other causes, say one-third more - 15 feet.

Total head from the bottom of the reservoir or from the commencement of the high pressure pipes - 45 feet.

To this must be added 20 feet for the depths of the reservoir and filter, giving the fountain head a total elevation of 65 feet above the highest streets.

The source from which it is proposed to take the supply of water is a stream which is formed by the conflux of three mountain streams at a place about 12 miles from Yokohama, called Kawaiemura, and which falls into the sea about half way between Yokohama and Kanagawa.

With a map of the district its drainage and or catchment basin could be discovered, and with proper returns of the fall of rain its discharge could be calculated at all seasons. With some knowledge of the geological formation of the country its mineral ingredients could be got at, and with some means of analyzation, its purity could be further tested. But all these we are without, and it only remains for us to get all the information possible regarding the water, by such observations and enquiries as are at present within our reach.

The necessary height of the fountain head, viz.: 65 feet, can be got at about the distance of 6 miles from Yokohama, at a place called Buko-jie-n ura.

On 27 and 28 February last, after four or five months dry weather, the flow of water in the stream by three separate gaugings was as follows:-

1st at the rate of .... 3,047,040 Gallons per day.
2nd ,, ,, ... 3,490,560 ,, ,,
3rd ,, ,, ... 3,831,840 ,, ,,

Making an average of 3,456,480 Gallons per day, and from information received on the spot from several persons who had been thirty and forty years there, this quantity is seldom less. Two years ago, and ten years ago, after on each occasion long dry summers, the stream was reduced to about one tenth of its present size, but it only lasted so for about thirty days, and this only occurred on these two occasions in the recollection of the oldest men. In wet weather the stream has sometimes six feet depth of water in it. One objection to taking water from this stream is that it is occasionally used for the purpose of irrigating the rice fields in its neighbourhood, several dams with sluices having been formed to lead the water into separate channels for that purpose when required. It will be seen, however, that less than one tenth of the present flow, or 300,000 Gallons, is all that is required to supply Yokohama, and though on the two occasions mentioned above, that was nearly all the water in the stream, they are of such rare occurrence that no extensive provision need be made for them. If it should be found necessary to make such provision, it could be done in two ways, viz.: by an impounding reservoir to collect a large store of water in the wet season, or by diverting another stream which exists about six miles beyond Ka-wai-e-mura into this, the two combined giving an abundant supply. The cost of neither of these works is allowed for in the following estimate, because they are certainly not required for the wants of Yokohama, and it is hardly probable that on further investigation they will be found to be required at all.

Quality of Water - That the water is collected from a partially cultivated, partially waste area, and flows for about six miles through a cultivated district is all that can be known of it, without elaborate investigations, requiring much time and attention. The impurities it is likely to collect are the pollution from the manure of the cultivated ground, decayed earthy matter, and mineral matter in suspension or solution. From the high esteem in which the water is held in the district for drinking purposes, it is probable that none of these exist in it to any great extent. The process of filtration to which it is proposed to subject it will be very efficacious in ridding it of a great portion of what impurities may exist.

By passing it through layers of fine sand, coarse sand, shells, fine gravel, and coarse gravel, the organic and earthy matters, as well as the mineral matter in suspension, may be separated, almost entirely, leaving only the chemical impurities, such as gasses, or matter in solution to cause any serious apprehension. These as far as can be judged by superficial observation are not in large quantities, but specimens of the water should be laid before analytical Chemists to be tested before these proposed works are carried out.

WORKS AT FOUNTAIN HEAD

Weir - A weir must be formed across the stream, at the point at which it is proposed to abstract the water of such a height that only the surplus water, over and above the quantity required, shall fall over it, behind it being the sluice through which the water gets admission to the filters.

Embankments - As in floods the stream overflows its banks, embankments will require to be formed to protect the filters, reservoir and other works from the flood waters.

Filters - The general construction of filters in England varies very slightly. Taking one recently constructed, viz., that for the new Chelsea Water works as an approved sample, the filtering material in it is as follows:-

1st - A layer of fine sand, 2 feet 6 in. deep.
2nd - Coarse Sand, 6 inches deep.
3rd - A layer of Shells, 6 inches deep.
4th - A fine gravel, 4 inches deep.
5th - A Coarse gravel, 2 feet deep.

Total depth, 5 feet 10 inches. The water being admitted on top of the first layer of fine sand percolates through the different layers and finds egress through a series of perforated pipes laid in the bottom. This filter is known to pass 700 gallons in 24 hours per square yard of area, therefore $428^1/_3$ square yards will be required for the 300,000 gallons to be supplied to Yokohama. But to this should be added an allowance for any extra demand of water, or to make up for loss of time during the cleansing of the filters or other contingency - say 530 square yards or a size of 80 feet by 60 feet, and a filtering capacity of 371,000 gallons for 24 hours.

Service Reservoir - This should be of sufficient size to contain three or four days supply, so that a good quantity may be at hand in case of fire, also to allow time for cleansing the filters, or for repairing works &c. Four days supply is 1,200,000 gallons, and to contain this the reservoir will require to be 102 feet long, by 100 feet wide, by 13 feet deep. All recently constructed service Reservoirs are covered, and this is considered beneficial, as it keeps the water in them at a uniform temperature and free from vegetation; an open reservoir on the other hand, with concrete sides and bottom, might be constructed at about one half the cost, and with careful watching and cleansing might be kept tolerably clean, but the water in it would be exposed to the heat of the sun and the action of the air and light. In this climate, notwithstanding its excess in cost, a covered reservoir may be considered essential, and its cost is allowed for in the estimate which follows:-

PIPING

The line of piping should follow the valley in which the stream flows and enter Yokohama beside the new road now in course of formation at the foot of Nungi-hill. To convey the water at high pressure from the reservoir to

Yokohama, and distribute it through the streets, no piping can with safety be used except cast iron, jointed in the usual way with lead. A pipe nine inches in diameter with a head or fall of 45 feet, and a length of six miles will discharge 389,700 gallons per day, and this size of pipe will be required to carry the total volume of water. When, however, Nungi, Yoshida, and the Native town of Yokohama has been supplied, the size of the pipe might be reduced, a quantity of 180,000 gallons per day having already been carried to its destination, leaving only 120,000 gallons to be further conveyed. For this a pipe of six inches diameter will suffice, its carrying capacity being 141,480 gallons per day, while three-inch pipes capable of discharging 24,930 gallons per day would be a sufficient size for most of the branch streets. This, however, is only the size of pipes required for the domestic supply and to make provision for fire these sizes must no doubt be increased. A nine-inch pipe discharging 389,700 gallons per day, or $27^1/_3$ gallons per minute, will throw jets equal in volume to three or four ordinary Yokohama fire engines, and this together with the supply now existing may be considered as sufficient. Nine-inch pipes should therefore be laid in all the principal thoroughfares, while in the smaller streets six-inch pipes discharging 98¼ gallons per minute would suffice, three-inch pipes being to supply solitary houses or very unimportant streets. These pipes should be laid all the way 3 feet 6 inches below the surface, to keep them out of the reach of frost or other disturbing influences.

DISTRIBUTION OF THE SUPPLY

Hydrants constructed so that fire hose could be coupled on to them should be placed in every street at a distance, say of 100 yards apart. In the Native town, Water Columns of cast iron fitted with keys may be placed at the corners of streets or in other convenient places, from which water could be drawn when required the key being given only to those who pay the water rate, as is done in many towns in England at the present time. The manner in which the water should be distributed among the foreign residents, will be a matter for future arrangement, and need not be discussed here.

ESTIMATE

The following is a liberal Estimate of the cost of works, which there is little probability will be exceeded.

FOR BRINGING WATER TO YOKOHAMA WORKS AT FOUNTAIN HEAD

Weir across River—40 feet long built of stone	$3,000
Embankment—To protect works from floods, say 300 yards at $6	1,800
Filters—Sides and Bottom to be concrete, and including necessary sluices for regulating the supply of water, 530 square yards at $15	7,950
Service Reservoir—To be built with stone or brick, with arched roof and including the cost of sluices, capacity 1,200,000 gallons at $25 per thousand gallons	30,000
Keeper's dwellings—&c., say	1,000
	$43,750

Piping—Nine-inch cast iron pipe 4½ miles long, weight
at 116 lbs. per yard 410 tons at $48 19,680
Freight on 410 tons at $13 5,330
Making 2,640 joints for do at 30 cents 792
Digging trench for pipes 3 feet 6 inches deep, filling
in ditto and making good surface 7,920 yards at 30
cents 2,376
Extra price for carrying pipe across streams, roads &c.,
say 1,000   29,178

                                                            $72,920
Add 10 per cent for contingencies   7,292

                              Total $80,220

FOR THE DISTRIBUTION OF WATER

Nine-inch pipe—In all the Main thoroughfares eight miles
at 116 lbs. per yard, 729 tons at $48 34,992
Freight 729 tons at $13 9,477
Making 4,700 joints for ditto at 30 cents 1,410
Digging trench for ditto, filling in and making good
surface 14,080 yards 30 cents 4,224
Extra cost of carrying pipe across stream &c. ... 1,000

                                                            51,103

Six-inch piping in the smaller thoroughfares ten miles at
78 lbs. per yards, 613 tons at $48 29,424
Freight of 613 tons at $13 7,969
Making 5,866 joints for ditto at 30 cents 1,760
Digging trench for pipes filling in and making good
surface 17,600 yards at 30 cents 5,280   44,433
Three-inch piping, for unimportant streets, &c., eight
miles at 28 lbs. per yards 66 tons at $50 3,300
Freight of 66 tons at $13 858
Making 1,760 joints for ditto at 20 cents 352
Digging trench for pipes, filling in and making good
surfaces, 2,280 yards at 20 cents 1,056   5,566

Hydrant in streets, every 100 yards, say 300 at $10 ...   3,000
Public Water Columns say 150 at $10 7,500

                                                            111,602
Add 10 per cent for contingencies 11,160

                    Total   122,762

TOTALS

For bringing water to Yokohama   80,220
For the distribution of the Water   122,762
Engineering Fees, superintendence and other expenses,
say   7,000

                                                            209,980

ESTIMATED REVENUE

Here as in England the Water rate should bear some proportion to the value of property, this being a matter for future arrangement. If the following average rates were charged, viz.:-

Native Houses,	5,059 at $4 each per Annum	...	...	22,236
Chinese ,,	200 at $6 ,,	,,	...	... 1,200
European ,,	250 at $7 ,,	,,	...	... 1,750

	23,186 (sic)
Less cost of Repairs, Salaries &c., say	2,000
	21,186

The amount realised would give a dividend of 10 per cent on the expended capital per annum, the rates being on the average with the present population.

Natives	$1.07 for inhabitant	... ...	per Annum
Chinese	$1.00 ,,	... ...	,,
Europeans	$2.91 ,,	... ...	,,

The low rate for Chinese is on account of their being so many of them merely domestics in European houses and therefore not chargeable. As the population increases, so of course will the revenue.

It may be mentioned for information that in England two compulsory water rates are levied, one being as an equivalent for Water used in household purposes, the other as an equivalent for the advantages it affords in the case of fire, for watering streets, for flushing drains or other public purpose.

The domestic rate varies in different towns - from 1/6 to 9d. per pound sterling on the annual value or rental of property.

The public Rate from 6d. to 1d. per pound on the same.

Engineers Establishment,
Benten, 5th March 1870

First published in THE JAPAN WEEKLY MAIL, 20 September 1873

LIGHTING

ILLUMINATION by means of Petroleum or Paraffine Oil is a subject which has been exciting considerable interest for some few years back. The results both in point of economy and photogenic power which have been attained by those who have brought out some of the lamps recently introduced shew that it is a material which will sooner or later eclipse all other oils, if it does not eventually supercede coal gas. In England, where natural petroleum does not exist, a kind of bituminous shale has been discovered in various localities, which on being distilled produces an oil of the greatest value. This crude oil is so treated as to separate its different component parts, and from it is produced the petroleum spirit, or what is commonly known as Naptha or Benzoline, and which is the highly inflammable and dangerous compound of the crude oil; the burning oil which may be more or less pure and inflammable according to the purpose for which it is required and the price to be paid for it: the lubricating oil which is the heavy portion of the distillate with all the lighter and combustible materal extracted from it: the paraffine, a white fatty substance now largely used in the manufacture of candles, matches, &c.; and tar which is the refuse from the distilling processes. The extraction of all these highly useful substances from a material, which is simply an aluminous deposit, and resembles ordinary clay stone, with the greatest nicety and precision and so as to form a most lucrative industry, is one of the greatest modern achievements of chemical science and manufacturing skill. The petroleum found in large quantities in America is of nearly the same nature as the crude oil got from the distillation of the shale in England, except that there seems to be generally less paraffine in it, and in some oils none at all. And in consequence of this less trouble is taken in America than in England to extract that valuable substance.

In Japan, in the provinces of Echingo and Shinano, there are found large quantities of petroleum, and it has been seen in the neighbourhood of Niigata, oozing out of the earth and running down a water course to the sea. The Government have recently granted a concession to a native company to work the petroleum springs in these districts, but it will be long before it will be able to produce an oil which will compete with the kerosine imported to this country from America. The company, which has called itself the "Petroleum Oil Company" and has its head quarters at Yedo, has sunk a well in Shinano three feet in diameter and from 400 to 500 feet deep. This well goes through various strata, and probably below the coal measures, but at the bottom comes upon a dark clay through which the petroleum exudes in large quantities. They raise it by means of a rope and buckets and manage to get up in this rude way as much as 16,000 gallons per month. The crude oil is described as of about the same colour as ink and the same density as common oil. The only process to which it is subjected by the native company, at present, is a rude attempt at distillation by boiling the oil in an iron pot which has a pipe leading out of it to carry off and condense the steam. With these means they get about 45 per cent or 7,200 gallons of oil, and 55 per cent or 8,800 gallons of tar per month. The spirit is not extracted from the oil nor is it in any other way purified. The whole of it is sold to natives, in the province of Shinano, to burn in lamps, at the price of ¾ boo per sho, or about 47 cents per gallon. The price in Yokohama would be 1 boo per sho or 63 cents per gallon. The tar is thrown away as useless. Specimens of the oil were given to the writer to test, and from the above account of the treatment it received it may easily be discerned that he found it to be utterly useless for illuminating purposes. It seems to be peculiar in so far that it contains very little spirit. It was heated up to 180° deg. Fahr. without giving off any inflammable vapour, whereas, what are considered safe burning oils in England, flash or become dangerous when heated to 130 degrees. Its specific gravity also was found to be 851, water being 1,000, whereas safe oils are generally about as low as 810 or 815. The Japanese oil is, therefore, an extremely heavy and very safe oil, but it is so impure, that in a few minutes the refuse from it clogs the wicks, and a black, tarry, substance keeps constantly running down the burner. Burning it in one of the most approved lamps for Petroleum, it only gives a strength of light equal to 9.5 candles, whereas the American Kerosine usually sold here gave a light equal to 19.76 candles.

It is said, however, that the Petroleum Company have ordered the machinery from America necessary for the proper purification of the crude oil, and have also engaged the services of several Americans to superintend its working. In this case, and if the work of purification is done with integrity and honesty, and the distribution of the machinery is so arranged as to enable the work to be done with economy, there need be little doubt than an excellent and cheap burning oil could be produced.

It is only very lately that any considerable attention was given to the best means of burning petroleum. The common flat wick lamp has, it is true, been for many years in use and still answers its purpose very well, but Paraffine oil has always been looked upon as a species of liquid gunpowder which it was more or less dangerous to have in any ways close proximity, and any improved methods of consuming it did not meet with much attention.

The manufacture of the oil has, however, now attained such perfection and its properties have been so investigated by scientific men that there no longer remains doubt regarding its properties or any prejudice against it, and several competitors have appeared in the field with new methods for consuming it to the best advantage.

It should not be understood, however, that all oil sold as Paraffine or Kerosine is safe; it is indeed, in Yokohama, quite the reverse. The writer had occasion to test several tins of American Kerosine lately, and the result was that in specimens taken from seven different cases marked Devoc of New York, two flashed or gave off an inflammable gas at below 90 degrees - three at 91 degrees, and two at a little over 100 degrees - while their specific gravities ranged from 785 to 788. This means that if, from the heat of the weather, from a fire or from other causes, this Kerosine gets heated to the temperatures above mentioned, it gives off a gas which will explode on any light being brought in contact with it. This of course is highly dangerous and should make people most careful as to what Kerosine they purchase. In England a safe oil is considered to be one with a flashing point of about 130 or 140 degrees temperature and a specific gravity of about 815, and a very simple test for all householders who are not sure of the quality of the oil they purchase is to have a small density metre which is easily procurable and only costs five shillings, and to see that no Kerosine they buy has a specific gravity of less than 810. They may then rest assured that they can use the oil they get with perfect security.

Of the various inventors who have brought forward improved lamps for the consumption of Petroleum oil the first who drew active attention to the matter was Captain Doty, an American. He patented a lamp which had for its specific object the burning of mineral oils in lighthouses. He has been so far successful that all the lighthouse authorities in England and the Continent agree that it will give nearly double the amount of light at less than one-half the expense, of the old colza oil lamp but while France, Sweden and other countries have adopted his burner, England refuses on the grounds that his patent is not a valid one and that his invention is not original. Captain Doty has, however, threatened to file an injunction in Chancery to prevent the English Government using mineral oils in lighthouses, without acknowledging his patent right, and there the matter has rested for the last three or four years. The Japanese Government last year came to the decision to adopt Captain Doty's lamp in all lighthouses and is now engaged in converting the different lights.

Captain Doty's lamp gives a light equal to 20 candles, while the ordinary lamp now in use in the lighthouses gives a light equal to 11½ candles, and at the same time Doty's lamp burns at the cost of oil in Japan only 9 cents worth per ten hours, while the common lamp burns 16 cents worth. As there are at present more than 200 burners in the Japanese lighthouse service the change will save the Government between five and six thousand dollars per annum.

Since Captain Doty introduced his burner several adaptations of it have been put forward, from none of which can such good results be got, but which all claim to be original inventions. Some of these are direct infringements of Doty's patent, while others it would be incorrect to condemn as such - though there is no doubt they are not inventions nor does their construction exhibit any original design. Among others that have followed in the wake of Doty is the loudly heralded lamp of Mr. Silber. This gentleman was fortunate enough to get access to the columns of the *Times* where his improvements on lamps were emblazoned in large type. Since then he has sold his patent rights to a Company dignified by the name of the "Silber Light Company" who in the most energetic way are giving the whole world the benefit of Mr. Silber's improvements, they having even turned their attention to Yokohama where the new lamps may now be had at the moderate price of from twelve to twenty dollars each.

The principles upon which all Kerosine lamps are made, and always have been is to give a free access of air to the flame, and by means of a cap or tube to cause an impingement of air upon both the outside and inside of the flame, so as to ensure a plentiful supply of oxygen. If the supply of oxygen is in excess, bending the tube or cap outwards from the flame will naturally decrease it; on the other hand, if the supply is too small, bending it inwards will increase it. This is all Mr. Silber has done and all he claims to have done; but surely this is no invention or worthy of being patented and heralded in the *Times* as a scientific accomplishment. The Silber lamps are merely an adaption of the common flat-wick lamp principle to a round wick. There is an abundance of holes for the access of air, and they are better made than lamps are generally as their price can warrant.

When in 1869 the *Elleray* was lost with the material for the Japan lights, the writer was obliged to hit upon some device to show temporary lights upon the coast, and he got a number of lamps constructed in Yokohama, from his own design, with round wicks and to burn Kerosine. These lamps, on comparing them with the "Silber lamp," are identical in every way. On trying them with the photometer they have precisely the same photogenic power, and while Silber's lamp burns one gallon of oil in 73 hours, these lamps burn one gallon in 80 hours. Further, a number of lamps used in America for the fronts of locomotives were procured from New York for the same purpose, and the principle of their construction is also identical with Silber's lamp.

The actual power of the different lamps, all having wicks of the same size, is as follows:

Doty's lamp	20	candles
Silber's lamp	16½	,,
Common flat wick lamp	10	,, (sic)

It will thus be seen that while Silber's lamp is more than ½ better than the common lamp, Doty's lamp is ¼ better than Silber's; and although Doty's is only at present used for lighthouse purposes it is quite easy to adapt it for domestic purposes.

At the ordinary prices of Kerosine in Japan the cost of the consumption of oil in these lamps burning for six hours each night is as follows:-

Doty's lamp	5½ cents.	
Silber's lamp	6	,,
Common Flat-Wick lamp	4¾	,, (sic)

Lamp for lamp, therefore, the common flat-wick lamp is considerably the cheapest, but as its power is so small it may not be, in some circumstances, the most economical to use.

The relative cost of burning them for 6 hours each night per candle power of light which they emit is:

Doty's lamp	.27 cents.	
Silber's lamp	.36	,,
Common Flat Wick lamp	.47	,, (sic)

Therefore when a great strength of light is required the Flat-wick lamp is the most expensive and Doty's the cheapest. With regard to the safety of the lamps a good deal depends on the material of which they are made as copper or brass conducts heat far more rapidly than porcelain or glass. The more brilliant the flame and the better the light the greater the heat that will proceed from it, and therefore a lamp giving a strong flame should have the reservoir containing the oil as far apart from it as possible. In Silber's lamp after 6 hours burning the oil in the reservoir rose from a temperature of 74 deg. to 86 deg., while in a commonly constructed flat-wick lamp it rose to 83 degrees. Both lamps are therefore safe to use with good oil, but with oil which flashes at 90 degrees there is doubtless some danger in using either.

All credit should be given to those persons who make improvements of this nature, and no one can question the great advantages in having a lamp which will consume perfectly and not emit smoke or disagreeable gases; but in Silber's case the alteration on the old lamp is so trifling, the old principle is unaltered, and lamps of precisely the same construction have been so often constructed before and the same results obtained from them, that to herald these lamps as the results of scientific investigation bears the impress of quackery, and to patent the slight and immaterial alterations proposed and carried out by Mr. Silber makes us very suspicious of the good judgment of those Commissioners whose duty it is to grant patents.

It should be understood that in the above remarks regarding Silber's improvements reference is only made to his house lamp with a circular wick of the ordinary argand size. He has patented various other improvements for different kinds of lamps which the writer is acquainted with but which are not touched on in this paper.

198

Two other companies have lately been started to work inventions for the use of Petroleum, but on an entirely different principle to those above mentioned. One is the "Air Gas Company," and the principle on which they proceed is very simple. By transmitting common air through the lighter Petroleum oils in which a gum is dissolved, it becomes so impregnated with inflammable particles that it becomes a species of gas and burns with great brilliancy. But unfortunately this company was doomed to disappointment, as it was found that, on the passage of air so impregnated through pipes, it became condensed and the inflammable particles returned to their original liquid state.

To remedy this another company started called the "New Gas Company." They eject a jet of steam upon molten metal; the steam becoming decomposed forms a hydrogen gas, and this they then, in the same way as the "Air Gas Company", pass through petroleum so forming a hydro-carbon gas which gives a very pure light and which is said to cost only 1s. 8d. per 1,000 cubic feet.

The great influence and power of the large gas companies in England will cause a delay in the introduction of any such innovation as this, but in Japan, where the erection of gas works is only beginning, such improvements should be fully considered.

First published in THE JAPAN WEEKLY MAIL, 25 September 1875

LETTERS TO THE EDITOR

THE ENGINEERING COLLEGE

Imperial College of Engineering, Tokio, 23rd September 1875.

SIR, - In your paper of last week I observe an article on Public Works, signed 'R.H.B.' Into the question of the management of Public Works, I will not enter at present, but I hope you will allow me a short space to notice a few of R.H.B.'s remarks on Engineering Education in general, and on this College in particular.

As far as I can gather from the article, the only fault which is found with the programme of this College is, that too much time is taken up with theoretical studies to the neglect of the practical part of the students' education. If this were the unanimous opinion of men who are thoroughly well qualified to judge of such a question, I would not have the slightest hesitation in modifying the course to suit their views, but I have received letters from men who stand high in the profession, and without exception they agree with the arrangement which I have made. If any alteration is suggested, it is to make some of the theoretical subjects more complete, but not one of my correspondents proposes that any of these subjects should be curtailed.

For instance: Mr Lewis Gordon, who is as distinguished in practice as in theory, says, 'I have read the whole of the programme - it is wonderfully complete and I see nothing too ambitious in the Technical courses.' Mr John Anderson, late Superintendent of Woolwich Arsenal, - a man who rose from the ranks of working men, and therefore likely to know the relative value of theory and practice - suggested that the technical course in mechanics be extended.

I might give you similar opinions from other men quite as qualified to give advice, but I think the following extract from the *Engineer* embodies all that the others have said upon the subject.

'As an example of what a Government engineering college should be, we cannot do better than cite the institution at Tokio in Japan. The Japanese Government are in just the same position as that of India. They want engineers, and they have established a college to supply them.

'The calendars of this college for 1873 and 1874 lie before us, and they contain a great deal that is eminently suggestive. The Imperial College of Engineering was established at Tokio in 1873, under the orders of the Minister of Public Works, with a view to the education of engineers for service in the Department of Public Works.

'All the students are Japanese and all the professors are British. The Principal is Mr Henry Dyer, C.E., M.A. University of Glasgow, while the Professors of Natural Philosophy, Mathematics, Chemistry, Drawing and English Literature are all men of high attainments.

'At the end of the calendar for 1874, examples of the examination papers are given, and although they are easy enough, it reflects no small credit on students who acquired their information through the medium of a foreign language - English - that they appear on the whole with much credit to themselves and their instructors.

'The principal reason we refer to this College here, is, to show the admirable manner in which it is proposed to combine theoretical with practical instruction.

'The course of training will extend over six years. During the first four years, six months of each year will be spent at college, and six months in the practice of that particular branch which the student may select. The last two years of the course will be spent wholly in practical work. By this alternation of theory and practice the students will be able during each working half-year to make practical application of the principles acquired in the previous half-year. The system of instruction will be partly what is usually called professorial, and partly tutorial, consisting in the delivery of lectures, and in direction and assistance being given to the students in their work.

'Can anything be better than this, or more likely to produce the class of men that is wanted? We regret that we have not space to describe more minutely an institution which has much to recommend it. We regret still more that English youths have no such facilities for learning their profession as those afforded by the Japanese Government.'

In drawing out the course for this College, the first thing to decide was the greatest length of time that could be spared for the students' education.

Taking into consideration the pressing wants of the country and the well known anxiety of the Japanese for some practical results, I felt that if we extended the course beyond six years, it might lead to dissatisfaction. For the first two or three sessions the standard of examination will be comparatively low, but I have not the slightest doubt that at the end of the sixth year, we shall be able to turn out men who will prove useful to the country.

The length of the course fixed, the next thing was to determine the relative time to be spent at theory and practice. The above extract from the *Engineer*, shows how I arranged it, and yet 'R.H.B.' says that he cannot resist the conclusion that too much importance is given in the Engineering College in Yedo to theory.

If I had proposed to take up all the time in theoretical studies, I might have been accused of giving too much prominence to theory, but with the present arrangement I refuse to admit the justness of 'R.H.B.'s' criticism, the fact being that I know of no College in existence where so many facilities are given for practical work.

Perhaps it may surprise 'R.H.B.' when I say that I had previously quoted to my colleagues some of the extracts from the report of the Institution of Civil Engineers given by him towards the end of his article, in order to justify my proposal not to attempt too high a theoretical standard for a few years, as I was impressed with the necessity of turning out practical men in the shortest possible time. It is evident from 'R.H.B.'s' article that he is quite ignorant of the facilities we have for practical work in connection with this College. Does he know that we have well equipped laboratories where what may be called experimental engineering may be successfully carried out? Does he know that we are forming a museum, where we are collecting models of machines and engineering works, specimens of materials and manufactured products, all of which the students will have opportunities of examining? Does he know that in connection with the College we have one of the largest (in a short time it will be the largest) mechanical engineering establishments in Japan, where those students who wish to become mechanical engineers will have every opportunity of learning the details of their profession? Even the Civil Engineers will spend a great part of their time here, as 'R.H.B.' need hardly be informed that now-a-days a man who is a good mechanical engineer has nearly learnt the practice of his profession as a civil engineer.

In the other departments special facilities will be given for learning the practice of their profession, so that at the end of their sixth year, there is every prospect of the students being able to exercise their judgment, instead of being bound hand and foot to a book of formulae and a copy.

To those who pass a satisfactory examination at the end of the sixth year, I proposed to give the degree of 'Master of Engineering'. The examination will take place in the principles and practice of the work on which the students have been engaged. Strictly speaking it is a mistake to speak of an examination in the practice of engineering; it should rather be called an examination in certain things pertaining to that practice.

No college can undertake to examine in the practice of any art; the proficiency of the students must be certified by the engineer under whom they serve, and who has observed their aptitude for such work. The Calendar says, 'The position of the student in the service will be determined by the final examination at the end of each course, as well as by his general aptitude for business.'

'R.H.B.' jumps to the conclusion that all the students will be immediately placed in charge of the execution of works on their own responsibility. Nothing could be further from what is intended.

In some cases where the students have spent a good deal of time on some special kind of work, they may be appointed to take charge of such work, but as a rule they will be under the supervision of some foreign engineer until they have obtained sufficient experience to undertake the work on their own responsibility; but, as I said, the general aptitude or inaptitude of the students for business will be taken into account, and if found necessary they may be appointed to very subordinate positions.

'R.H.B.' says the possession of a degree 'implies a capability to design and conduct the execution of engineering work.' I would ask him to look over the Report of the Institution of Civil Engineers, and see if the diploma given by any college in Europe pretends to do anything of the sort. As a rule the diploma is given when the student leaves college, before he has spent any time in practice, and it simply certifies that the holder is in possession of such knowledge as will enable him, when he attempts practical work, to carry it out in an intelligent manner.

I admit there is something anomalous in the name 'Master of Engineering,' but the same anomaly exists in other professions - for instance, 'Master of Surgery' is given to a medical student when leaving college, but it is never considered to mean that its holder is thorough master of all the principles and practice of surgery, it merely signifies that he is in possession of such knowledge as will enable him by practice to master all the details of his profession.

'R.H.B.' has quoted Professor Fleeming Jenkin to show that the course for Civil Engineers is too extended. If he turns to page 8 of the same report he will find the programme of Professor Jenkin's class, and will see that the list of subjects is practically the same as what I have proposed, with this difference, that his is more advanced. To show you, however, that Jenkin's course is not above Japanese students, I know of two who attended his class in Edinburgh, and they were able to take places considerably higher than the average students. I should advise 'R.H.B.' to study this report of the Institution of Civil

202

Engineers, and he will see that what I have proposed is the merest elements of what is attempted on the Continent. When he finds us setting such examination papers as are printed at page 123 of the above mentioned report, I will give him permission to say that we give too much attention to theory.

In speaking of the Cooper's Hill College 'R.H.B.' says, 'he is inclined to the opinion that such institutions, by lessening the time which young men have to devote to the practical part of the profession, and by destroying their taste for it, will more or less tend to lower the high reputation in which English engineers stand in all parts of the world.' It seems almost a waste of time to argue with one who holds such opinions, opposed as they are to those of every right thinking man. If they had been uttered two hundred years ago, they might have been forgiven, · but in the present day when every small village has its Mechanics' Institute, and every large town its Technical College, they can only proceed from one who has some special reasons for holding them.

'R.H.B.' quotes George Stephenson, Brindley, Telford and Smeaton as examples of what can be done without theoretical knowledge. Every one admits that these men by their mechanical genius were able to accomplish wonders, but none lamented more than themselves their ignorance of mechanical laws, and they never missed an opportunity of improving themselves in the theory of their work.

Does 'R.H.B.' not know how much George Stephenson regretted his deficient education - how, when a boy, he saved his three pence a week so that he might learn penmanship and arithmetic - and that his determination to procure a first class education for his son - the want of which he himself felt - was what spurred his ambition, and no doubt contributed to his success?

Brindley, when designing any new work, was in the habit of retiring to bed, and there arranging his plans, a system which 'R.H.B.' might advise to be tried with the Japanese. If Brindley had even known the elements of arithmetic and mechanics, he would have been able to have done double the work with less trouble.

Telford and Smeaton seized every opportunity of improving their defective education, and none would have been more willing than they to admit the practical utility of a thorough grounding in the principles of their work.

It seems almost unnecessary to enlarge further on this part of the subject, but I will shortly mention another engineer, celebrated for building light-houses, Robert Stevenson of Edinburgh, of whom 'R.H.B.' may have heard. If so, is he aware that this Robert Stevenson, when building the Cumbrae Light-house, spent every spare moment he had in going to Glasgow to obtain instruction in mathematics and mechanics at the Andersonian University? Does he know that when engaged on a light-house on the Pentland Skerries in Orkney, he came to Edinburgh University in the winter and studied mathematics, natural and moral philosophy, chemistry, natural history, logic and agriculture? I have no doubt 'R.H.B.' thinks this time was sadly mis-spent, or at least would have been spent to greater advantage in practical work; but Stevenson thought otherwise, and I am very much inclined to agree with him.

I would have 'R.H.B.' remember that other branches of engineering are taken up in this College besides civil. Would he also apply his arguments to the training of mechanical engineers? If so, then I have small hope of future progress in that department. Would Watt have made any of his discoveries if he had not previously studied the laws of heat and principles of mechanics? I very much doubt it. Even the small improvement that has taken place in the steam-engine since his time has been the result of theoretical investigation, backed up of course by practical experience.

203

To-day I found a striking example of the value of a knowledge of theoretical mechanics to a mechanical engineer or mill-wright. I was taken by a Japanese official to see what was considered a wonderful machine, which was intended to revolutionize the world and supply any amount of power out of a small pond of water. The apparatus consisted of a water-wheel, which worked on one side a set of pumps, and on the other a set of the ordinary Japanese stampers for cleaning rice. The action of the machine was intended to be as follows:- the water-wheel was turned by hand sufficient to effect a few strokes of the pumps, - these discharged the water into a vessel at a considerable height, and this again fell upon the water-wheel in the ordinary way, - which in turn supplied power to drive the stampers.

The water passed from the wheel into the pond to be used over again, thus intending to realise the Perpetual Motion, the dream of school-boys and philosophers of all ages. The practice of this man was excellent, but his theory was deficient: he was considered a mechanical genius by his neighbours, but now I think he is a wiser, as well as a sadder man. How much trouble and expense would he have been saved, if he had known the elements of the principle of the Conservation of Energy. Does 'R.H.B.' think that this man's education would have been complete if he had been able to read a book of formulae? If so, I would recommend him to study the Patent Office list, where he will find examples of frequent occurrence in which this principle of the Conservation of Energy and others equally important, are continually violated. Professor Rankine remarks:- 'Such men too often spend their money, waste their lives, and, it may be, lose their reason in the vain pursuit of visionary inventions of which a moderate amount of theoretical knowledge would be sufficient to demonstrate the fallacy; and, for want of such knowledge, many a man, who might have been a useful and happy member of society, becomes a being, than whom it would be hard to find anything more miserable.'

If 'R.H.B.' looks a little further on in Fleeming Jenkin's lecture which he has already quoted, he will find the following:-

'The younger men in factories have to design machinery frequently of one type, but nevertheless varying so much from year to year, that a sound knowledge of mathematics, mechanics, and physics, is of the greatest importance to them; consequently, as draughtsmen or designers, we find foreigners employed all over the country. And, à priori, I would rather engage a foreigner to carry out my ideas in designing a new machine than a young Englishman of equal standing, especially as most of the foreigners complete their education by working in a shop for a short period.'

It is impossible to follow his logic when he tries to prove that what is required for mechanical is not so essential to civil engineers - in fact, he is inconsistent when he says further on 'I wish my engineering pupils to owe quite as much to Professors Kelland, Playfair, and Tait as to me.'

I cannot agree with 'R.H.B.' when he says that in Japan nothing new in the way of engineering is likely to be required. If by new, he means new in principle, he is probably correct, but I think that Japan offers innumerable opportunities for new arrangements and designs, both in civil and mechanical engineering.

If these designs are made by a mere copyist, we are likely to see even more wonderful productions than the specimen I have mentioned above, where each part may be correct when taken by itself, but the general arrangement contrary to some of the most ordinary laws of mechanics and physics.

'R.H.B.' remarks that 'in the ordinary practice of the profession few questions of theory arise which have not already been worked out by such men as Fairbairn, Rankine and others, and have been reduced to formulae by them - the application of which is simple in the extreme.' He is rather unfortunate in the names he selects, as Fairbairn was not a theoretical man, he was entirely practical. The little theory found in his books was not given by himself, but by a mathematical friend, whose assistance he always acknowledges.

Fairbairn's favourite theme at meetings of Mechanics' Institutes was his own defective education, and the necessity for the study of theoretical subjects to obtain success in practice. Rankine again, though his works are master-pieces of the kind, has not reduced his formulae to such simplicity that the 'veriest tyro in mathematics' can understand them, as 'R.H.B.' would have us believe. I should not like to trust the work designed by such a tyro.

I have far exceeded the limits which I intended for this letter, and will conclude by inviting 'R.H.B.' to Yedo, that I may show him what we have done and what we propose to do, that thus he may be in a position to treat his subject with a greater amount of justice.

With this I send two copies of our Calendar for the current year, one of which you may present to 'R.H.B.'

I am,
Yours very truly,
HENRY DYER.

PUBLIC WORKS

Yokohama, 26th September, 1875.

SIR, - I find it is necessary to say just a few words in reference to Mr Dyer's remarks in your last issue. He has addressed himself to statements which I never made, and to arguments the purport of which he has entirely misapprehended.

It might have been expected that a reply, shewing a want of appreciation of the point of an argument, which fails to answer it on its merits, but rushes wildly to an extreme which nothing I have said can warrant, and in which a number of irrelevant illustrations are made use of, would have emanated from a mind accustomed to deal with romances, and which has not been disciplined to the close and solid reasoning which accompanies mathematical training; but, from Mr Dyer, something might reasonably have been looked for which would have displayed a more logical train of thought.

Nothing is further from my intention than to depreciate the value of theoretical knowledge. For the intelligent treatment of any subject it is all essential. For the purposes of invention, for the design of any structure of which examples do not exist, for the proper application of examples, and also for the ordinary and every day practice of the profession of civil engineering, a knowledge of mathematics and general physics, in a greater or less degree, is necessary, and, as I have already stated, the more an engineer knows of these, the more useful he will probably be.

But I also said, and this I am fully prepared to maintain, that, in order to train a man who will be capable of efficiently designing or carrying into execution, any ordinary engineering work, theory may safely be made subservient to practical experience, in cases where it is difficult or impossible to attain a knowledge in both of these. I also showed how difficult it has been found to combine these two branches of knowledge in one person in England.

It is possible for a man with a good sound general education, and with a series of years of intelligent observation in the discharge of duties connected with the actual construction of work, to be an extremely useful practical engineer, but it is perfectly impossible for any man coming direct from any college, to either design or satisfactorily conduct works.

It is surely too bad in Mr Dyer to ground his reply to such an argument as this, upon an assumption that I object to an education in scientific subjects for engineers altogether, and to go out of his way to make use of catchpennys, which, he must be aware, can have little influence; such as enquiring whether I was aware of George Stephenson saving three pence a week to learn penmanship, of Smeaton's lamenting his ignorance of mechanical laws, of Brindley's retiring to bed to think over his plans. I probably knew all this before the subject of engineering ever entered Mr Dyer's head, and I mentioned these men's names, not with the object of advocating illiterateness, nor as instances of what engineers of the present day should be, but as examples of what it is possible for practical knowledge to accomplish without theory, while at the same time I gave instances of what theory had endeavoured to accomplish in India without practical experience.

The issue to which I would desire to direct Mr Dyer's attention is whether, in training a man for the practical duties of the engineering profession, it is not advisable to limit his theoretical studies to those only which the highest authorities affirm to be absolutely essential to the ordinary practice of the profession, and in this way to give a more enlarged opportunity to him for the acquirement of practical knowledge.

I ventured upon a remark in my former paper that such institutions as Cooper's Hill College in London, by lessening the time young men have to devote to the practical part of the profession, and by destroying their taste for it, will more or less tend to lower the present high standard of the engineering profession. To this Mr Dyer replies that 'it seems almost a waste of time to argue with one who holds such opinions, opposed as they are to those of every right thinking man' - 'they can only proceed from one who has some special reasons for holding them.' I cannot help thinking that it is a pity Mr Dyer should have permitted himself to write in this way, because such expressions justify a personal retort which, were I to use it, would probably give irresistible force to my arguments. But I refrain, because nothing is further from my intention than to show any hostility to Mr Dyer or his labours.

A justification of my remarks is not far to seek. In a recent leading article in the *Times*, the conduct, in India, of the young gentlemen who had been sent out there from Cooper's Hill College is very sharply commented on. These gentlemen are said to have formed the impression, that, having gone through and successfully accomplished their course of study in London, their mission in life has been accomplished. They have attained a position which their former hard studies have gained for them, and they feel no necessity for further exertion. They do not work, as an Engineer requires to work; such labour is probably considered derogatory and beneath the attention of men possessed of an extensive knowledge which they feel that they have but little occasion to apply. They have made themselves conspicuous by bullying the natives, by insubordination and by other behaviour which, at all events, shews some defects in their training.

Hoping that Mr Dyer will form, in future, a juster appreciation of my arguments,

I am, Sir,
Your obedient Servant,
R.H.B.

ENGINEERING COLLEGES

October 6th, 1875

SIR, - I do not intend to say one word about this College; I only mean to give you an opportunity of showing 'R.H.B.' that the '*proof*' of his letter - as founded on an article in the *Times* - referring to Cooper's Hill College, was utterly without foundation and therefore his whole argument falls to the ground.

For this purpose I enclose an article from the *Spectator* of July 31st and another from the *Overland Mail* of July 30th showing plainly that the writer in the *Times* made a mistake in the article quoted by 'R.H.B.' which mistake Lord Salisbury, the Secretary for India, took the first opportunity of rectifying.

The *Overland Mail* also gives his Lordship's speech at Cooper's Hill, which you can read for yourself and see if it justifies the remarks of the *Times*.

I also enclose you an article which appeared in the *Engineer* of July 30th which will give you the opinions of engineers generally with regard to that institution, that thus you may be in a position to compare their opinions with those held by 'R.H.B.'

Lastly, I send you an article from the *Engineer* of August 13th, which will show you what the Russians are doing in the way of Technical Education.

If you think these articles will interest your readers or enlighten 'R.H.B,' I hope you will make such extracts as you may see fit.

I am,
Very truly Yours,
HENRY DYER

[We have selected the following extracts from those sent by Mr Dyer. The former refers to the Engineering College at Moscow; the latter to the Cooper's Hill College.-[Ed. J.W.M.]

First published in THE JAPAN WEEKLY MAIL, 22 January 1876

NOTES TAKEN DURING A VISIT TO OKINAWA SHIMA - LOOCHOO ISLANDS

BY R. HENRY BRUNTON, ESQ

Read before the Asiatic Society of Japan, on the 19th January, 1876

AMONG the first papers read before the Asiatic Society was one by Mr E. Satow, of H.B.M. Legation, upon the Loochoo Islands. The following notes may be considered as affording a little supplementary information, which was procured during a couple of days' visit to these Islands.

We left Kagoshima in the *S.S. Thabor*, on the evening of the 9th December, and proceeded so as to pass to the westward of the chain of islands which extends in a south-westerly direction from Cape Chichakoff (Satanomisaki) to the north end of Formosa - the Loochoo islands forming part of the chain. On the evening of the 10th we passed, at a distance of about thirty miles, Oshima, which is one of the Loochooan group, and is inhabited by people of similar kindred, but which is under the authority of the Kagoshima *ken* and does not form a part of the territory of the King of Loochoo. It was on this island that a sugar refinery was erected by the Prince of Satsuma with the assistance of the Messrs Glover of Nagasaki, in 1867. This did not prove successful financially and, I understand, the machinery has since been removed to Ozaka. Oshima possesses an excellent harbour which is sheltered in every direction and the trade between it and Kagoshima, in certain seasons, is very considerable.

We arrived off Nafa on the afternoon of the 11th December. During the voyage we experienced N.W. winds of no great violence - and this wind continued with us more or less until our return. The monsoon whose direction is N.E. on the China coast, is therefore changed in direction in this locality.

Nafa is on the western shore of Okinawa Shima; it is a town of probably five or six thousand inhabitants, and is supposed to be the port for the capital of the island, which is situated about three miles inland. The distance from Kagoshima to Nafa by the track we came is about 400 miles. The harbour at Nafa can hardly be considered fit for vessels to make any extensive use of. There is only a very slight indentation in the coast line, which there runs about N.E. and S.W., and the only protection to vessels is afforded by coral reefs which partially surround a basin. Some of these come quite to the surface of the water and their position may generally be distinguished by the surf breaking on them; others, however, are at distances of 3 feet and 6 feet and 12 feet below the surface, and nothing is visible which marks the position of these latter. And, as they seem to exist at the entrances to the anchorage as well as in the anchorage itself they form a most formidable danger. The strong N.W. wind which blew during one day that the vessel lay at Nafa cut off all communication between the ship and the shore, and a heavy swell came rolling in to the anchorage which rendered our position a most uncomfortable and, to some extent, a precarious one. We knew that we were surrounded by perfectly precipitous coral reefs, some of which were within a few hundred yards of us, and had any fracture occurred to the moorings which held the ship as she plunged and heaved, we should, in a few minutes, in all probability, have been driven against one of these. In a creek on the southern side of the town there is a depth of water of from two or three fathoms. Small steamers might find shelter here, and three or four large junks were lying in it at the time of our visit. But for vessels of over 300 or 400 tons, it is not available, and Nafa cannot in any way be considered suitable for a commercial port.

When we arrived at Nafa, the fact of our being in a country in which the people were very different from those whom we had just left, became at once apparent. Several boats came off to the ship, but instead of being the tidy-looking, swift, and picturesque craft which we are accustomed to in Japan, they were canoes, made out of one log of wood, and very similar in shape and appearance to those at Aden. They were propelled by two men, one in the bow and another in the stern, by means of paddles. The only other boats which we saw were square clumsy looking boyes which were also propelled by paddles. The sea-going junks, however, which make voyages to Foochow and Kagoshima are strongly constructed, and are of the same shape and build as the ordinary Chinese junk. They are decorated with an eye on each bow and by a red ball on a white ground on the stern. This, we were informed, however, was merely a decoration, and in no way an emblem that the vessels were under Japanese colours.

The town of Nafa is built on a piece of level ground adjoining the sea coast, while the capital, Shiuri, is built on a series of small hills. The latter is a straggling scattered town, and it would be very difficult to form any estimate of its population. From the summits of some of the hills good views can be obtained of the surrounding country, which appeared to be in the highest state of cultivation. And from its gentle undulations, with small streams winding through the valleys, its rich herbage and avenue of trees, it afforded a very close resemblance to some phases of English scenery.

The streets in the towns present a most desolate appearance. On each side of these is a blank stone wall of about 10 or 12 feet high, with openings in them here and there sufficiently wide to admit of access to the houses which are behind. Every house is surrounded by a wall, and from the street they convey the impression of being prisons rather than ordinary dwellings.

The streets are also paved with blocks of stone. These have very irregular surfaces which render walking over them most uncomfortable. It was observed that in front of the dwellings of those of high station, or in front of temples and other places of importance, that the roads were laid perfectly smooth, with broken stones bound by clay in much the same system as is used on macadamised roads. The people therefore, are perfectly aware of what smooth roads are, and it is to be regretted that they did not more largely adopt these. The high road from Nafa to Shiuri is 30 or 40 feet wide and is lined with trees on each side. It is laid throughout its entire width and for the whole length of the road with these blocks of stone. This represents an immense amount of labour and, while not convenient for pedestrians, it has the advantage of being everlasting.

The houses of the well-to-do classes are situated in a yard which is surrounded by a wall 10 feet high, as has been already mentioned. They are similar to the ordinary Japanese houses with raised floors laid with mats, and sliding screens of paper. They are built of wood and present no peculiar differences from the Japanese style of construction. The roofs are laid with tiles, which, however, are quite different in shape to the Japanese tiles. Over the joint between two concave tiles, a convex one is laid, and these are all semi-circular in cross section. The tiles are made at Nafa and are red in colour. They appeared of good quality. The houses of the poorer classes are of a very primitive character. The roof is covered by a thick thatch, and is supported by four corner uprights about five feet high. The walls consist of sheets of a species of netting made of small bamboo, which contain between them a thickness of about six inches of straw. This encloses the whole sides of the house, a width of about two feet being left in one side as an entrance. There is no flooring in the houses of any description, and there is generally laid over the mud inside a mat on which the inmates lie or sit. We found that a pig was generally attached to each of these houses and that pork is very largely consumed by the inhabitants of the island. In each house also there is a weaving loom, and the dresses of the people are all woven by themselves in their own houses.

There are no shops in Loochoo, and when anything is required to be purchased it may be brought by the dealer to the house of the buyer. There is, however, in each town a market-place where various commodities are exposed for sale. These principally consist of the general food of the people, which is sweet potatoes, pork and a few fish. There were large quantities of Japanese tea which we were told the Loochooans were very fond of, and which they drank after the fashion of the Japanese. Satsuma tobacco was also observed in some quantity, and several bundles of English cotton twist. The market stalls are all presided over by women, who are evidently entrusted with the commercial operations of the island. The only money in use is copper cash, but the natives did not refuse the silver of some of our party who purchased a few things. Some of the women in the market were young, but the great proportion were elderly. The practice of tatooing the backs of the hands of the women was to be seen here; the younger ones had a few marks only, while the hands of the elderly ones were covered down to the nails. That the Loochooan married woman are kept in such seclusion as is related by Mr Satow, namely, that they are not allowed to have any communication whatever with the small portion of the outer world, may be true of the higher classes. We had no opportunity of testing this, but that it cannot be correct as regards the lower classes there can be little doubt. We observed numbers of women of all ages engaged in all manner of occupations, going about and conversing with as much freedom and self-assurance as is customary in any part of the world, and it is only reasonable to suppose some of these must have been married.

Loochoo has been visited by various foreign vessels at different periods. H.M. ships *Alceste* and *Lyra*, then on a cruise in Chinese waters, came to Nafa in the year 1815. And Captain Basil Hall has described the Island in his narrative of the voyage of these vessels. They remained at Nafa for some months and were refitted there, and the officers and crews experienced the greatest courtesy and civility from the natives. Commodore Perry also called here on his way to Japan with the U.S. squadron, but his experiences of the Loochooans were not so favourable as those of Captain Basil Hall.

A little distance out of Nafa under some fine old fir trees are quite a number of graves of Europeans. Each grave has placed over it a block of stone work, about the same size as an ordinary grave and three feet high. On the top of this there are, on most of the graves, stones let into the masonry, which have inscriptions cut on them. From these we observed that two or three Catholic priests had been buried there, as also four men who had belonged to the American squadron. One inscription was over the grave of an English sailor of the *Alceste*, and it bore testimony to the good feeling existing between the English and the inhabitants at that time. It mentioned that the memorial had been erected by the 'King and inhabitants of this most hospitable island.'

The people are of a timid and most inoffensive nature, and all our experiences of them shew them to be kindly disposed to strangers. Their treatment of two survivors from the wreck of an English brig which came ashore on the island some years ago, was so considerate and so highly appreciated by the home authorities, that a gun vessel was despatched to offer the thanks of the English Government, and to present to the king a gold watch in recognition of the kindness shown to them by him and his people.

On the sides of most of the small hills in the neighbourhood of Nafa are the tombs of the inhabitants. They are built of stone into the sides of the hills. Their top resembles a horse-shoe in shape, and in front there is the opening into the interior which is built and cemented up so as to be air-tight. The roof is made of plaster and is flat, and the appearance of a number of these tombs on the sides of the different hills has a very picturesque effect. The method of burial is to leave the corpse in the tomb for about three years until it is entirely decayed, the tomb is then opened, the bones are collected and kept in an urn as a relic by the family. While walking in the neighbourhood of the tombs we observed one open. A temporary mat shed was erected in front of it and sounds of the most violent grief proceeded from the interior. This, we understood, was on the occasion of the re-opening of the tomb for dis-interment when the relatives meet to witness the last rite performed upon the remains of the deceased. Mr Satow mentions that this ceremony has been discontinued and that the people are buried in the same way as Japanese are; but I have reasons for thinking that this is only partially true, and that it is still adhered to by certain classes of the population.

There is abundant evidence that the whole island of Okinawa is of coral formation, and it, in all probability, affords an interesting example of a coral island which has been, since the formation of the coral, subjected to volcanic upheaval. On a hill about two hundred feet above the level of the sea the exposed rock was distinctly of coral and, in many places further inshore, coral was observed. The stone used in the buildings, walls and also in paving the roads was also undoubtedly coral, but it had, of course, lost to a great extent its characteristic appearance by wear and exposure. The whole country is low and undulating, no hill being above 400 or 500 feet high, so that the upheaval to which it has probably been subjected has not been of an extremely violent nature. Coral reefs rising to about the surface of the water surround it on every side. These generally enclose a central space of deep water. The passage through these at Nafa, as is customary in other coral islands, is opposite the mouth of a fresh water stream. The island is subjected to frequent shocks of earthquake, showing further that it is not yet far removed from volcanic action.

The climate is one which is of a sufficiently genial character to allow the vegetation to be green throughout the whole year. In the winter months some cold days are experienced, but snow or ice is unknown. The thermometer, during our stay in the month of December, was as high as 73° Fah. in the shade, and the sun was sufficiently powerful to necessitate our return to thinner apparel, sun-helmets and umbrellas.

The product which we observed as chiefly cultivated was the sweet potato. It is the principal food of the people, patches of paddy land were seen here and there, but they were not of any great extent. The rice is grown at various seasons of the year as it suits the convenience of the farmers. We saw some which had just been planted and which was expected to be reaped in March or April. The climate will admit of two crops being produced in the year, but the people, with sensible prudence, do not tax the productive properties of the land to so great an extent. Many groves of sugar-cane were observed.

211

Oranges of a peculiar aromatic flavour grow on the island, but not in large quantities, and fruit of any kind appears scarce. The sago palm is cultivated in large quantities, and the sides of all the hills which are not otherwise occupied are covered with it. The whole island is in the highest state of cultivation, and in this respect will bear a favourable comparison with any part of Japan. The farming implements in use seemed to be precisely similar to those used in Japan.

There are many ponies on the island. They are from 10 to 10½ hands high and are well-shaped little animals. Some of our party who rode on them gave excellent accounts of their spirit and paces. They resemble the Manila breed of pony, or may possibly be a cross between it and the China breed.

A few cocoa nut trees were seen but they do not bear fruit. Small quantities of tea and tobacco are also grown.

The trees on the island are all of small size, and wood is not plentiful. A few teak trees were seen, but the natives do not consider that they produce a valuable timber and consequently pay them no attention. From our own observation the wood in them would be small and full of knots. A hard timber named *Komon* and a soft wood named *Fuchitsuba*, are grown on the island and were observed in the temples and other erections. But a great part of the wood in use is the ordinary Japanese wood which is imported from the Kagoshima.

Owing to the scarcity of wood stone has entered much more extensively into the building operations of the people than it has in Japan. The execution of the mason work is also infinitely superior to anything to be seen in Japan. The Loochooans seemed to have grasped the principle of giving strength to mason work by friction between the beds of adjacent stones. The joints are therefore made with truth, and although the stones are generally of the most diverse shapes, they fit into each other with perfect accuracy, unlike Japanese masonry in which the stones are only kept in place by a bearing on each other of a few inches wide on their outer edge. The walls which surround the dwellings and which line the streets are all built in this way. The material in them is small, and in a few cases we observed that lime mortar had been used in the joints.

All the bridges, on the roads, which we observed were built of stone. The openings in them are spanned by means of arches in the form of an ellipse. So far as the eye can judge they are almost perfectly elliptical. This form of arch is also to be seen over all gateways and other openings. There are also a few arches of the form of a segment of a circle, but the elliptical arch is, by far, the most common. Some of these are of exceedingly small rise and are ingeniously constructed. They are however faulty in so far that the arch stones are placed lengthways along the entrados of the arch, instead of the joints radiating from the centre. They are therefore but ill calculated to bear the strains which come upon them. The bridges have well constructed piers and abutments and are furnished with stone parapets. The inside of these are ornamented with elaborate carvings and designs, which are chiefly of Chinese origin. The Loochooans admit having received their notions of building from China, and those familiar with that country will probably see great resemblances to Chinese erections in all that exists in Loochoo; but there are also, without doubt, evidences of some departure from Chinese methods and of original ideas on the part of these islanders.

The castle which is the residence of the King is situated in about the centre of Shiuri and on an eminence about 500 feet above the level of the sea. The dwellings, in which the king resides, are placed near the summit of the eminence. They are of the ordinary type of wooden buildings and are of considerable extent. They are built in a square, enclosing a court about seventy yards wide. This court is laid out in paths with different coloured tiles. Opposite the entrance is the largest building, which was shut up at the time of our visit - on the left and right being smaller buildings which are used as the residences of the court officials, as reception chambers, and for other purposes. Surrounding these buildings but at lower elevations, on the sides of the hill, are very extensive revetment walls, some of which are 60 or 80 feet high and 14 or 15 feet thick. These walls, which must sustain an enormous pressure from the earth behind them are built in plan, in the form of a series of inverted arches. This seems an ingenious and excellent expedient for assisting the strength of such walls, and a principle something similar to it is adopted in large retaining walls in Europe. The pointed part of the wall, from which two arches spring, is ornamented by a peculiar and graceful curve. Cactus in profusion grows along the tops of the walls and they are covered on the outside with various creepers. To enter the castle it is necessary to pass through three gateways of very heavy mason work, the openings through which are spanned by elliptical arches. The castle cannot be called fortified, though its position and the existence of the high walls already mentioned would render it safe against capture except from an attack furnished with modern appliances. The object in constructing it has no doubt been to make it a stronghold capable of resisting any enemy known to the people. The king seems to take but little active part in the Government of the country and had not been visible to any stranger for some years.

The country altogether presents a strange admixture of Chinese with Japanese ways and customs. The inscriptions which are to be seen on the various monumental stones placed in the streets are written in Chinese. Many are quotations from Confucius and other Chinese classics, while others go so far as to represent the country as part of China. The principles of building have been partly borrowed from China. The tombs and manner of burial are to some extent after the Chinese fashion, and the general appearance of the towns gives evidence of the existence of a close intimacy with China. But the language on the other hand is Japanese. It is not precisely similar to the Japanese now spoken, but it is believed to be as nearly as possible similar to the language spoken some centuries ago, many words of which have become obsolete. While a great part of what was said by the Loochooans was understood by the Japanese, many words used by them were recognized as belonging to the old Japanese vernacular but which are now never heard. The most frequent intercourse with Japan has led to a familiarity with Japanese ways and the use of Japanese produce.

There were no Chinese on the island, and we heard of only one Japanese merchant who was resident there. There are three or four Japanese officials living at Nafa who have been probably sent there for the purposes of the Government.

A tax of 8,200 *kokus* of rice is levied by the Japanese government from the Loochooans. It is paid in sugar, that being the most valuable produce of the Island. Certain articles are also sent to China each year, but these consist merely of complimentary offerings and small presents. Communication is kept up between Loochoo and Foochow by means of large junks.

The Island is between four and five hundred square miles in area, and it is said to have a population of 150,000. This is at the rate of three hundred to the square mile; a thickness of population so great as to be hardly conceivable.

The people are burdened with the maintenance of a large class of idlers who like the *Samurai* in Japan live upon hereditary privileges granted them by the government. These men are called *Daimios* and may be seen lounging about in every street. The lower classes are an industrious, docile, timid, but extremely civil set of people. Their education is confined to Chinese, and only one book of Chinese classics is taught in the schools or, in fact, is known among them. They have little, or no communication with the outside world. They produce on the island what is sufficient for their own wants and would probably be best satisfied in being left alone. But this is not to be their fate. The Japanese Government have taken over the active control of the country. It has been formed into a *Han*. A Mitsu Bishi mail steamer now visits it once every alternate month, and the Loochooans have probably experienced the last of that quiet and peaceful retirement which the geographical position of their country has heretofore afforded them, and for which their natures seem so well adapted.

APPENDIX 5

CONSTRUCTIVE ART IN JAPAN
[PAPER 1]

BY R. HENRY BRUNTON, ESQ.
M.I.C.E., F.R.G.S., F.G.S

[Read before the Asiatic Society of Japan, on the 22nd December, 1873]

THE ACCOUNTS of Japan which at the present time are generally spread throughout Europe, are so exaggerated that both the natural beauties and wealth of the country as well as its genuine condition and the progress which it has made are greatly over-estimated by those who have not had an opportunity of visiting the country and of judging of them for themselves. Every one, therefore, who comes to Japan is led to expect too much, and there are few who on arrival do not experience feelings of disappointment. And it is probable that nothing develops these feelings more fully than the absence of those artificial improvements which are generally met with in all civilised countries. The dwellings of the people are of mean appearance, and are generally without ornament or adornment of any kind. They are built in a temporary and unsubstantial manner, and are to a great extent wanting in the comforts which are ordinary in all European houses. The streets in the principal towns, as well as the country roads, are rutted, uneven and perfectly untended; and although gravel is sometimes used, the roads are generally merely formed of the earth or clay through which they pass. There is almost an entire absence of drainage, and the refuse water from the houses is allowed to spread itself over the streets. The rainwater has no means of egress, and lies in pools until it has time to sink into the earth or is evaporated.

It is further impossible to repress a feeling of disappointment when we turn to the religious monuments of the country. The temples are stately, they are generally exquisitely ornamented, and are certainly built in a more stable and substantial manner than the other erections around them. But there is so great a sameness about them that it seems as if the original designer had made a groove so deep that all the intellectual power of the Japanese could not raise their architects out of it.

That earthquakes are prevalent throughout the whole of Japan is a fact which, in the minds of many, has affected the whole system of building in Japan, and has prevented the development of the native talent for construction. This is looked upon as sufficient reason for the absence of stone erections or buildings of solidity and durability. But if earthquakes have exercised this influence over the Japanese mind, the people have been influenced by false premises; as I think that to imagine that slight buildings, such as are seen here, are the best calculated to withstand an earthquake shock is an error of the most palpable kind. Now that foreigners have introduced a different system of building, the present Japanese have no hesitation in adopting it, and edifices of any size or material are now erected with their approval. No objection is ever made on account of earthquakes, and on these grounds I am of opinion that at all events the present race have not that dread of earthquakes which would lead them to eschew solid constructions, and we must seek at some other source the reasons for the want of progress in the art of building.

The whole country is subject to earthquakes, and there is hardly an island or a province of Japan that has not at some time or other suffered from their effects. Through the courtesy of certain Japanese officials I have been put in possession of some information, which I have every reason to believe to be correct, regarding the destructive earthquakes which have occurred. From this I gather that the country is becoming more and more liable to them, and that they have steadily increased in number during the last few centuries. Thus there was but one destructive earthquake in the 5th century, which happened in the year 415; none other is known to have occurred till the 10th century; one more occurred in the 11th century, another in the 13th, two in the 16th, 10 in the 17th; 13 in the 18th and 15 in what has already passed of the 19th century. The average of this century therefore has been one destructive earthquake in every five years, while 800 years ago there was but one in 50 years. The following is a list of the most destructive which have occurred throughout the country.

In the 5th year of the reign of the 20th Emperor - in the year 415 - a destructive earthquake happened.

Another in the year 937, another in the year 1021, and another in the year 1292, which was felt worst at Kamakura.

One felt worst at Tsuruga and Totomi in the year 1588.

One which destroyed many houses and took many lives at Kioto and Fushimi in the year 1595.

One at Yedo which destroyed the Castle and many Daimios' residences in the year 1647.

Another at Yedo which knocked down many houses and killed a great number of people; in the year 1649.

One in the province of Iyo which brought down the retaining walls of the Castle of Matsuyama and destroyed many houses in Uwajima; in the year 1649.

One severely felt throughout the eight provinces surrounding Yedo in the year 1650.

'One which partially destroyed the Castle of the Mikado at Kioto and ruined the castle of Nijo in the year 1661.

One felt in the province of Echigo in the year 1661.

One felt in the Island of Yezo in the year 1662.

One which again partially destroyed the castle of Nijo at Kioto when the shocks lasted for eight hours; in the year 1662.

One felt at Nikko in the year 1682.

One felt in Dewa in the year 1693.

One felt throughout the eight provinces surrounding Yedo. Walls of outside and inside moats of Castle of Yedo destroyed. Felt very severely at Odawara, where many houses were destroyed and numbers of people killed. Tidal waves also broke along the coast at the same time, and caused enormous destruction. The road leading through the Hakone pass was closed up by alteration in the surface of the earth; in the year 1702.

One severely felt in Yedo in the year 1715.

One felt throughout the 15 provinces surrounding Kioto - when many parts of the earth opened up - and enormous tidal waves occurred in the year 1707.

One felt severely in the neighbourhood of Fuji no yama. At this time, which was on the 22nd of the 11th month, fire burst from a place called Moto hashiri kuchi at the base of Fuji no yama - there was a fearful noise like thunder, and a black gritty sand was thrown into the air which caused darkness to come over the whole surrounding country. Even in Yedo lanterns were used in the day-time. During the night of the 22nd this continued, but on the morning of the 23rd the sky was seen. On the 25th darkness again came on, black sand fell like rain, and it only cleared up again on the 28th. A small mountain rose up on the side of Fuji no yama at this time which has been called Ho-yei-zan, from the period in which the occurrence took place, which was in the year 1707.

One felt at Nagasaki, when there were more than 80 shocks in one day and night; in the year 1725.

One felt in the province of Echigo, which occurred during a heavy storm of wind and rain. The earth is said to have opened up and belched forth water, so that the plains were like rivers, and men, horses, cattle and all the animals in the neighbourhood were drowned; in the year 1726.

One felt at Kioto in the year 1751.

One felt in the province of Echigo, when the earth trembled 30 times in 10 hours, a hill was cracked, the earth opened and 16,300 lives were lost; in the year 1751.

One felt at Awomori, when the falling houses took fire and caused the death of a great many people; in the year 1766.

One felt in Yedo in the year 1771.

One felt in Yedo during the same year 1771.

One felt in Yedo in the year 1782.

Frequent severe earthquakes in Yedo in the year 1789.

One felt in Dewa, when both the hills and the plains were cracked and the earth opened up, in the year 1804.

One felt in the Island of Sado, when there were constant shocks for three days from the 1st of the 1st month and from the 18th of the 6th month in the year 1810.

One felt in the vicinity of Yedo, but worst at Kanagawa and Hodogaya, where many houses were destroyed; in the year 1811.

One felt in the vicinity of Kioto in the year 1818.

One felt in Oshiu and Yezo, when the earth shook more than 150 times; in the year 1821.

Frequent severe earthquakes at Yedo in the spring of the year 1824.

Frequent severe earthquakes at Yedo in the autumn of the same year.

One felt in the province of Echigo in the year 1827.

One felt at Kioto, when the Mikado's residence, many of the temples and the Castle of Nijo were destroyed. The earthquakes commenced on the 2nd of the 7th month, they partially discontinued on the 20th of the 8th month, but were not entirely quiet until the following year; in the year 1829.

One felt in the vicinity of Fuji no yama in the year 1833.

One felt in Sendai, when the castle was destroyed and great destruction was caused by tidal waves; in the year 1833.

One felt in the province of Shinano, which destroyed many temples and houses, numbering in all about 5,000; 700 people were killed and 1,460 wounded. The earth opened and swallowed 16 houses; in the year 1846.

One felt at Kioto and Osaka; in the year 1851.

Frequent earthquakes throughout the eight provinces surrounding Yedo, which were also felt at Kioto and in the Islands of Shikoku and Kiushiu. The earth was not quiet for one year; in the year 1854.

The most recent which has happened was most severely felt at Yedo, where the trembling of the earth continued for one month and gave 80 severe shocks. Many houses were knocked down, their timbers took fire and conflagrations commenced at 45 different places. About 120,000 lives were supposed to have been lost. This occurred in the year 1855.

Those parts of Japan most subject to earthquakes are, strange to say, the vicinities of the two capitals. Thus out of the 43 severe earthquakes which have taken place during the last 600 years, nine have occurred at Kioto and 13 at Yedo. The province of Echigo is next in number, and has had four earthquakes. Yedo has been visited twice, as also Dewa and the neighbourhood of Fuji no yama - while Nagasaki, Sado, Sendai, etc., have only suffered from one disturbance.

But, while the country, as is abundantly shewn above, is liable to very severe and increasingly numerous earthquakes, the system of construction in the buildings has not been well devised to withstand such visitations. The more solidity and weight in a building and the greater its inertia, the less liable it is to derangement from a sudden movement of its foundations; but at the same time, it is essential that the strength and connection of the materials in the walls are proportionate to their weight and mass. As a general principle preference should be given both on account of durability and stability to the adhesion of bricks or stone and mortar in a solid well built wall, over ordinary wooden buildings. It might be that a wooden erection could be constructed with its framework so tied and braced together as to render it almost perfectly secure against any earthquake, short of an upheaval or breach in the surface of the earth; but this would be an expensive, thriftless and impracticable style of construction. Whereas on the other hand, a stone erection need not be more than ordinarily massive to make it capable of resisting any shock not extraordinary violence. But in stone houses it is absolutely necessary that the masonry should be executed in a proper manner, the great point to which attention must be given being that a perfect bond is maintained throughout the entire building.

Mr. Mallet in his history of the Neapolitan earthquake of 1857 gives many proofs of the truth of this. He says: "When the masonry consisted of round lumpy quadrated ovoids of soft limestone, the whole dislocation occurred through the enormously thick ill-filled mortar joints, and almost all buildings thus formed fell together in the first movement in indistinguishable ruin." "Where the masonry was of the best class, and such as would be so recognized in England, the buildings thus constructed stood uninjured in the midst of chaotic ruin. Some examples of this will be found in the second part, none more striking than the Campanile of Atena, a square tower of 90 feet in height and 22 feet square at the base, in which there was not even a fissure while nearly all around was prostrate." "Indeed it was evident that had the towns generally been substantially and well built or rather the materials scientifically put together, very few buildings would have actually been shaken down even in those localities where the shocks were most violent. Thus the frightful loss of life and limb were as much to be attributed to the ignorance and imperfection displayed in the domestic architecture of the people, as to the unhappy natural condition of their country as regards earthquakes."

A very striking example of the advantage of solid construction over lightness and want of strength was seen not many years ago at Manila, when an earthquake levelled almost the entire town and left the stone lighthouse at the harbour, which is a column of masonry of great height, standing by itself perfectly unharmed. From the vast and handsome edifices which may be seen in most countries in Europe liable to earthquakes, we may conclude that their inhabitants have acknowledged the correctness of this principle, and it cannot therefore be urged on sound grounds that it is owing to the liability of Japan to earthquakes that its people have never desired or made an effort to build other than wooden houses or to make these of any but of the most flimsy description.

The general poverty of the people and their extremely simple habits may account for the simplicity of their dwellings, and as their habits become more refined and luxurious it is very probable that the internal comforts of their houses will also improve. Six hundred years ago the dwellings of the English were constructed in the roughest manner of wood and clay. The inmates ate and slept in one room and privacy was perfectly unknown. In the beginning of the 15th century the houses began to be divided into rooms and private apartments. Shortly afterwards glass windows and chimneys were introduced, and stone buildings were erected, the ruins of some of which are in existence at the present day. Gradually improvements were one by one effected, until the modern English residence was produced.

At present in Japanese houses there is a want of privacy, for although there are apartments, they are separated from one another by paper partitions, which accomplish their purpose only in name. There are no healthy or safe means of artificially heating the houses, and chimneys have never been adopted. There is an entire absence of glazing, and the light finds its way into the houses through the paper windows. These paper windows generally comprise a great part of the walls of the houses; and as they are very slightly made and do not shut closely up, the houses are extremely cold and unhealthy in winter. During six months of the year in the greater part of Japan the weather is such as to require properly shut up houses with good fires, and although during the other six months considerable heat prevails, it cannot be said that the style of building is at all suitable for the climate of the country.

The construction of the houses is of an extremely fragile and temporary nature. The structures consist of wooden uprights resting generally on rough round stones. These support the roof, the main beams of which are formed of very large timbers put in their place in their natural state, and without being squared or cleaned. The covering to the roof consists either of thatch, of tiles, or of shingles alone, and in putting these on the workmen are very expert. There are no diagonal struts between the uprights in the frame of the house, and no other means adopted to strengthen or stiffen it. The roof trusses are formed of one square frame built on top of another of a larger size until the apex is reached.

Thus, with its unnecessarily heavy roof and weak framework, it is a structure of all others the worst adapted to withstand a heavy earthquake shock. I should not forget to mention the fire-proof stores of the Japanese. These are buildings with a wooden framework of a better description, which is covered with sometimes as many as 50 coats of mud plaster, but generally with not more than 25 coats. They sometimes have a plaster roof and sometimes an ordinary tile roof. The plaster is of a thickness of from one to two feet, and the doors and window shutters are frames of wood covered with plaster in the same way. These stores, as is well known, have been found remarkably efficient in resisting fire.

On account of the simplicity of their construction and their general similarity, very little can be said regarding the temples of Japan in a paper such as this, which is devoted merely to a description of the art of building. The manner of their ornamentation and a history of their contents would form the subject for a separate and a very interesting paper. The foundations consist generally of square stones on which the uprights rest. These are of keyaki and are connected together at various intervals by longitudinal waling pieces. The roof is formed in a similar manner to the ordinary dwelling house roofs, but the wood in the beams is generally of keyaki and of great size. The roofs are generally thatched with the bark of the hinoki tree, or with a grass named kaya, which is put on to a thickness of, at times, three feet, in some instances they are covered with sheet copper and, in the case of the smaller temples, tiles are often used. The casing of the walls is thick keyaki planking on the outside and sometimes thinner hinoki planks as a lining on the inside. The outside is generally ornamented by panels of carved work illustrative of some legend or romance of the religion to which the temple was dedicated. The projecting ends of the beams of the roof have often some fantastic device carved on them, and are sometimes merely covered with copper to protect them from the effects of the weather. The joints of the various beams are also covered with copper.

The timbers used in the structure are joined together by mortices, scarfs or dovetails in such a way that metal fixings are seldom required and, with the exception of a few small nails, are but little used. But there is the same want of diagonal struts or ties in the framework of the temples as in the framework of the dwelling houses, and while the execution of the practical carpentry is generally excellent and the wood always of the best description, the manner of their construction is, in this respect, decidedly faulty. There are many temples in Japan from 200 to 300 years old, such as one at Shiba in Yedo which is 270 years old, and the wood used in them is still fresh and sound. A very fine modern specimen exists at Narita, about 30 miles to the north-east of Yedo, which is much thought of, and which was only built 18 years ago, but neither do the principles nor the details of its construction differ in any way from the ordinary specimens.

In some branches of carpentry the Japanese are very expert, and as their buildings are almost entirely of wood, the concentrated energy of the people seems to have been devoted to this branch of building. The neatness of their work is very noticeable, the joints of the timbers are made with the greatest nicety, and as paint is never used, these are exposed, and are so made an object of especial care. The frames of their paper windows are generally models of delicate workmanship, and the carved ornamentation in their houses or temples is generally beautifully executed. But when we come to the higher branches of carpentry, such as the arrangement of various beams so that they will be best adapted to bear the strains which are likely to come upon them, or a combination of timbers which will form a stiff, strong, and reliable structure, or the selection of the proper size of wood to stand the different strains which it will have to bear, then we find the Japanese very deficient. The carpenters do not seem to have any appreciation of the disposition of strains in any framework, and where enormous timbers are placed they may be found resting on and sustained by beams not one-quarter the size they should be. In their bridges the same incongruities may be observed; thus beams, which if properly fastened would form a tie and be a great support to the structure, may be observed secured in their places by wooden keys about one inch square, which are not much stronger than a match. The workmen, however, are very skilful in the use of their tools. They only require explicit and detailed directions, and they are then competent to execute any work in a very creditable manner. The woods generally used for building purposes in the southern parts of the country are not very varied.

There is a great variety of very excellent woods in the island of Yezo, but these have not yet been introduced into this part of Japan. Keyaki is the commonest hard wood and is, generally speaking, a very serviceable timber. If cut when ripe, and at the proper season, the good qualities will last for centuries, proof of which is shewn in the older temples in the country, but there are great varieties of quality and it requires a very practised eye to pick out the good from the bad. The exigencies of the people are such that, in the absence of any regulations to the contrary, they do not hesitate to cut the wood at all seasons, or when they receive an order for it. Wood full of sap is therefore as common in the market as seasoned wood, and perhaps it is not until after some years that the quality of timber purchased is made evident by the decomposed sap oozing out of it like a black tarry liquid. The fibres of the wood, very soon after this occurs, become rotten and the whole timber useless. Hinoki is the favourite soft wood of the Japanese, and is chiefly prized on account of the beauty of its grain and colour. It is also thought to be very lasting and is always used in erections which are intended to be durable. *Sugi* is a kind of cedar, and grows in large quantities throughout the whole of Japan. There are many qualities of Sugi, the best being almost as good in appearance as hinoki: it is, however, much cheaper. Sugi is principally used in the dwelling houses of the people which are only desired to be of an ordinary description. A cheaper wood which is used for more temporary erections is *Matsu*, a sort of pine. This wood is also used in bridges as, being a long fibred wood, it bears a considerable transverse strain, but it is by no means a durable wood. *Kuri* or horse chestnut is a very hard wood which does not grow to any size and is principally used for piles below water. A wood very much resembling ash, named *Kashi*, is used for boat's oars, handles of implements, etc. *Hiba* is very lasting under water, and is also used for piles. *Tsuga* is a kind of Hinoki but of very good quality. *Momi* is a cheap wood something resembling Matsu and used for the same purposes.

There are various other woods grown in this part of Japan, but the above are those most commonly used for building purposes.

The following are the names of the woods grown in the Island of Yezo - thirty-four in number, specimens of which I have received, and I have now the pleasure of presenting them to the Society. Sakura or common cherry tree, Shiki Sakura, a kind of cherry tree which is said to blossom in all seasons, Yanagi or Willow, Kada Sugi or Cedar, Kuwa or Mulberry tree, Niga-no-ki or Mulberry tree, Momi or Pine, Kurumi or Walnut, Yezo-matsu or Juniper, Kuri or Chestnut, Katsura, a sort of vine, Momiji or Maple, Kashiwa or Oak, Sugu-nara another kind of Oak, Ishi-nara another kind of Oak, Hannoki or Alder, Hachigo Hannoki another kind of Alder, Shuro a Palm tree, Ho or

Honobei, Yenju, Midzuki, Ouko, Aburagi, Tosen, Kisen Tani-chi-tamo, Aka-tamo, Nana kamado, Asada, Shiuku, Itaya, Gambi, Doro, Shina.

The art of building in stone, of brick-making, or an appreciation of the properties of lime has been very much neglected by the Japanese. Perhaps it would be too much to expect that the genius of the ancient Romans, to whom civilization is indebted for its present knowledge of building operations, should find its counterpart in Japan. Still if we consider that this country lays claim to a history of upwards of two thousand years, during the whole of which time it has been inhabited by the same intelligent race which at present occupy it, and if we compare the evidences of constructive ability to be seen in Japan with what may be seen in almost any other part of the civilized world, it is impossible to resist the conclusion that the subject has never received that attention to which it is justly entitled, and that in consequence there has been an utter want of progress in Japan in the art of building. The liability of the country to periodical and violent earthquake disturbances may possibly have had an influence in deterring the people from the use of stone, but, if so, I have already explained, I think it has been founded on false grounds.

The country is extremely well supplied with stone. Few districts of any extent are without it, and even with the native means of conveyance, stone might be procured in almost any town in Japan at a moderate price, if the methods employed in quarrying it were more perfect. Along the whole course of the Inland Sea the formation is igneous and granitic, and the stone of excellent quality. Many of the mountain ranges throughout the country are also composed of granite, and excellent quarries exist at Mount Chikuba, which is not more than 100 miles distant from Yedo, and to which there is inland water communication the whole way. The other stones fit for building purposes consist principally of hard unstratified clay stone and stones of volcanic formation. These are found in various localities and especially at many points on the sea coast. There is a soft sandstone largely used in the neighbourhood of Yokohama, brought from the Provinces of Sagami and Boshu, which is evidently quite a recent formation, and unfit for any building intended to be lasting. There is also a stone of white appearance much employed, but it is of little use except for the very questionable expedient adopted by foreigners here; which makes it take the place of tiles and plaster as the outside casing for the walls of their wooden houses. The only really serviceable stones at present used in this neighbourhood are got from Idzu, about 80 miles distant.

The stone erections which have been executed in Japan are very unimportant. On my making enquiries whether there were any stone houses in Yedo, I was informed that the only one was a house built 100 years ago by Nakagawa, then Governor of Osaka. It is constructed of granite brought from the neighbourhood of Osaka, but as it is only 12 feet by 9 feet and 10 feet high it is not a very imposing erection.

If we go back as far as the period of the Pelasgic architecture, which dates from 30 centuries ago, when the Pelasgi erected throughout Asia Minor and the whole south of Europe those wonderful specimens of their constructive skill which still exist, and if we compare their system of masonry with what may be seen in Japan at the present day, we can appreciate the want of progress made in this country. The walls of the Pelasgic erections were formed of stones of immense size put together without mortar. The stones when taken from the quarries were cut into irregular polygons and placed together in such a manner as to make the different faces of the geometrical figures which they employed coincide. This system of building resembles very closely what is

221

to be seen at the castle of Osaka, or at the moats and gateways of the Castle of Yedo. But while the Pelasgi themselves gradually improved and adopted the use of square stones laid on a flat bed, while in later years the ancient Romans gave a further impetus to the science and have left such specimens of their skill and knowledge of the properties of materials as their aqueducts and great roads, the Japanese have not moved, they still employ the same crude systems of building in stone, and are still ignorant of the most rudimentary principles of this branch of constructive art. The old Roman arch which marks an era in the history of building has no place in Japan. There certainly exist at Nagasaki, Kagoshima and in other places in the south, several specimens of semi-circular stone arches, but these were introduced by the Dutch residents and have never been largely adopted. In this also the Japanese have shewn a great want of appreciation of the art of building, and are behind their neighbours, the Chinese, in whose country I understand miles of stone arches may be seen, some of which are of almost incredibly large span.

Such stone work as is executed in Japan is put together perfectly dry, and it is an extraordinary circumstance connected with this subject that the people appear to be quite ignorant of the cementing properties of lime or of the use of lime mortar.

The use of mortar dates from the period of the invention of the Roman Arch some centuries before Christ, and was commonly used by the Romans of those days; but even up to the present day, after some years of education by foreigners, Japanese workmen will persist in laying stones on top of one another without any substance between them to fill up irregularities or to cement one to the other. Solidity in their masonry is not considered necessary, and the beds or joints of the stones are not made flat or even. The spaces between them are therefore large and are generally filled with pebbles, which are all that keep the stones in their proper places. Not being acquainted with the use of common mortar, it is unnecessary to say that they are also ignorant of the value of hydraulic mortar.

The Romans also taught us a lesson in regard to this which I am astonished to say has not been followed even by ourselves to the extent which it might have. They mixed the lime with Puzzolana or volcanic sand, which gave it the peculiar property of hardening under water; this mixed with certain proportions of gravel, formed concrete which, being thrown into the sea between moulds, in a short space became a solid and hard wall. Various moles or piers exist executed by the ancient Romans in this way. In England, where volcanic sand cannot be had, it has been discovered that a mixture of certain clays with lime has a similar and more perfect effect, and the mixtures so made are known as Portland or Roman Cement. In Japan large quantities of Puzzolana exist, and lime stone is also found in various localities, but I can learn of no instances where the mixture of the two was ever attempted. The principle of hydraulic cement is, however, known to the Japanese, and a substance which is formed by a mixture of lime and clay is often used by gardeners as a lining for fish ponds, and for other purposes, but the process of mixture is either defective or the materials used are not good, because although the cement hardens under water to some extent, it does not harden sufficiently, and it further cracks and falls to pieces when exposed to frost. Though acquainted with the principle, therefore, the Japanese seem to have been unable to bring it to any practical result. A lime plaster is made which is tolerably efficient, and is formed by mixing lime with boiled seaweed. But in plastering a house the first coat consists of mud generally procured from the bottom of some sluggish stream, the second coat of the same substance, this time mixed with sand presumably to harden it, and the lime plaster is then put on as the third coat, but so extremely thin that it is merely a veneer to the mud below it.

A curious system of building retaining walls, sea walls, or the face walls for any embankment or cutting, is so general throughout Japan that one is almost led to believe that the people had discovered some peculiar merit in it, although it is patently in contradiction to all our received notions of masonry. It consists of placing stones on one another which on their face are square or nearly so, but which are pyramidal in shape, and come to a point at their back. They rest at their faces on the thin ledge at the front of the stone and are supported at their backs by small stones loosely inserted, and the walls so built have generally a rubble backing about three or four feet thick. As a retaining wall or one which has to sustain a thrust of earth from behind, such a system of building is in utter defiance of all the principles of mechanics, because the stones are like wedges placed the wrong way: they have absolutely nothing to keep them in their places, and any thrust from behind must inevitably dislodge them. As a sea wall it may have this advantage, that a wave striking the stones from without acts like driving a wedge home, but it possesses this great defect that it does not afford solidity or strength, which is the great desideratum in any construction exposed to the force of waves. As a mere veneer on the banks of a canal or river to protect them from the action of the water, it may be efficient enough, but, if no more than this is required, an equally effectual and much cheaper method would be to line them with thin flags or wooden boarding.

The Hatobas in Yokohama, which have been broken up since they were erected by each heavy gale of wind that has occurred, were built in this way. The retaining walls of the creek in Yokohama, which were only built a year or two ago, and parts of which come down with every heavy rain, were also built in the same way, and it is so common, and the native quarrymen are so accustomed to cut out these peculiar pyramidal stones, that one of them can be bought at nearly one-half the price of a square stone of the same cubical contents. The intention or the advantages of this shape of stone I have never been able to discover, and although I have made enquiries of officials acquainted with the processes of Japanese building in all parts of Japan, I have never succeeded in getting a satisfactory reply.

This paper would not be complete unless I made some mention of the bronze images to be seen in various parts of Japan, principally because they are, without doubt, the most meritorious of all the attempts at construction which the Japanese have made. These stand out by themselves as evidences of a skill which it would be difficult to improve upon.

The mixing of the metals which compose bronze was practised in the earliest ages, and the casting of bronze images or statues dates from many centuries before the Christian era. Ancient coins as far back as the time of Alexander the Great were made of bronze, and, from an analysis which has been made of them, they have been found to contain from 17 to 6 parts of copper to one part of tin, with some other ingredients which it is not necessary to mention. Ornamental bronzes brought from Assyria have been found to contain 8 parts of copper to one part of tin. And the bronzes made in Europe of the present day consist generally of about the same proportion, viz., 8 parts of copper to one part of tin, and zinc or lead is sometimes added in quantities according to the purpose for which the alloy is to be used.

The Japanese bronzes differ in an extraordinary way from all these. From what I can gather the mixture generally consists of the following parts:-

To one part of gold there are added 8-9 parts of mercury, 33-65 parts of tin, and 1272 parts of copper.

There is therefore only 1 part of tin to 20-38 parts of copper, while the large quantities of gold and mercury, as far as I can discover, seem not to have been used by other people at all, and must add very much to the cost of the bronze.

The largest bronze image in Japan is at Nara, some distance to the eastward of Kioto. This idol was first cast in the 18th year of Tempei (in the year 743). It was twice destroyed during the time of wars in its neighbourhood, and the idol which at present exists was erected about 700 years ago. The casting of this idol was tried seven successive times before it was successfully accomplished, and about 3,000 tons of charcoal were used in the operation. The total weight of metal is about 450 tons and it consists of the following ingredients:-

Gold	....	500 lbs. avoirdupois
Tin	....	16,827 ,, ,,
Mercury	....	1,954 ,, ,,
Copper	....	986,080 ,, ,,
		1,005,361 lbs.

It is cast in pieces, and these pieces are joined together by a kind of solder which is called *handa-ro*, and which answers its purpose very satisfactorily. A few of the dimensions of the figure may be of interest.

Total height of figure	....	53.5 feet
Length of face	....	16 ,,
Width of face	....	9.5 ,,
Length of eye	....	3.9 ,,
Length of ears	....	8.5 ,,
Width of shoulders	....	28.7 ,,
On the head there are	....	966 curls.
Palm of hand	....	5.6 ft. long.
Middle finger	....	5 ,,

The image is surrounded by a glory or halo 78 feet in diameter, on which 16 images 8 feet long are cast.

There are two images standing in front of the larger idol, each of which is 25 feet high.

The whole is enclosed in a temple 290 feet by 170 feet, and 156 feet high, the roof of which is supported by 176 pillars.

The various pieces composing the image are not fitted together in a very finished manner, but the cement keeps the joints perfectly tight and close. The whole construction is one which shows great skill and original genius in the mixture of the metals and in the methods of casting them, and it is further one which will, no doubt, be a source of pride and gratification to the Japanese for many centuries to come.

In the beginning of this paper I referred to the conspicuous absence in this country of artificial improvements. These form, to a great extent, the work of the civil engineer, and it is interesting to examine to what extent the Japanese have mastered the various branches of science which are connected with that profession. In the means of internal communication the country is sadly deficient, and as these may be taken as the measure of a nation's advance in civilization, it seems remarkable that so little has been done by the present progressive race of Japanese to improve them. The roads throughout the country have not been formed with the intention of wheeled vehicles being used on them. Their surfaces are uneven and irregular, and little skill has been shewn in the choice of route so as to avoid hills or to get the best possible gradients.

There are many rivers which, if properly tended, would form excellent means of transport, but in some cases these have been neglected and in others treated in an erroneous manner. The Tonegawa, the largest river in Japan, has a bar across its mouth on which there is not sufficient water to allow the native junks to pass over it. Inside the bar there is a considerable depth of water, and the river is navigable for small craft for more than 100 miles. The Shinano-gawa, the second largest river in the country, has 6 feet of water on its bar, and there is little doubt that this might be deepened with ease were proper means taken to effect this. It has been allowed to break through its original confines until it is in some places two or three times its proper width, and is so dammed back by shallows that in floods the water overflows the banks and spreads over hundreds of square miles of rich cultivated country. For how many hundred years this natural process of washing away the banks and widening the river has been going on without check, or for how long it has been allowed to flood the adjacent lands, I am not in a position to say, but a step was recently taken with the avowed intention of remedying the latter evil, which, however, has proved unsuccessful. Instead of keeping such an enormous river, which is equal in volume to that of the Rhine, in the course which nature ordained for it, and taking the natural and more easy method of training its banks, regulating its width and inclination, and, if necessary, straightening its course, the Japanese conceived the idea of cutting another and separate channel to the sea for the purpose of carrying off the flood water - a great part of which has been already executed - but the works are now stopped. The design was erroneous in so far that the abstraction of the flood waters would probably result in a further shallowing of the natural course of the river, so entirely destroying its usefulness as a means of transport.

In bridge building the Japanese have a way of their own which has at least the merit of being quickly, easily, and cheaply accomplished. The piers generally consist of wooden piles driven a few feet into the bed of the streams. In some cases stone is used, and then it is cut to the same shape and of the same size as a wooden pile under the same circumstances would be. The platforms of the bridge are always of wood, and are generally constructed of longitudinal beams formed of a tree grown with such a bend as it may be desired to give the roadway. The bend is always considerable in Japanese bridges. The beams are laid 4 or 5 feet apart, and on top of them are laid cross planks which form the roadway. The span of each opening never exceeds 40 feet and generally is not more than 30 feet. One of the longest bridges in Yedo is the *Yei-tai Bashi*, which has 24 spans of 30 feet each.

The Japanese seem always to have been alive to the necessities for a plentiful and pure supply of fresh water. Yedo has had its water-works for many years, and the native town of Yokohama will also very soon be supplied with water in the same manner. The source of supply for both places is the River Tama-gawa, and the fountain-head is about 13 miles distant from each place. There is a dam across the river for the purpose of collecting the water into pipes, but there is no settling pond, filter, reservoir or other such appliance for purifying or storing the water as was used by the ancient Romans and is generally attached to water-works of the present day. The pipes are constructed of wood about 1 or 2 inches thick, and are made in the shape of a square trough, the joints being rendered tight by the insertion between them of a certain bark. The main pipes are from 1 foot to 2 feet square, and the smaller ones used for the distribution of the water are generally about 4 inches square. In the Yedo water-works the pipes are carried across valleys and streams on piles, but at Yokohama siphon pipes have been introduced. There appears to be some confusion in the Japanese mind in regard to the natural law that water always finds its own level. They appear to be cognizant of it so far, that they make allowances for the water rising in the siphon pipes and wells which they have adopted, but, on other hand, they do not appear entirely to have grasped the principle. In illustration of this, in Yedo there are placed five large wooden tanks at points where there are alterations in the inclination of the pipes. Thus, if they wished to supply a district higher than the level of the water main, instead of allowing the water to gravitate direct to that district they direct it first into one of these large boxes and allow it to rise there to the height which they desire, and then they carry it off from the box to the district requiring the supply. In the same way in the Yokohama water-works there are large boxes of a similar kind at each end of the siphons which carry the water under streams or other obstructions, so that instead of the water flowing direct through the pipe and along the siphon, it empties itself into the box at one end in the first place, the box then supplies the siphon, and the siphon empties itself into a box at the other end, from which the water proceeds along the main pipe. The adoption of these boxes must, I think, proceed from some misapprehension of natural laws, and I have been unable to discover any sufficient reason for them. The water is distributed through the towns in circular wells which are constructed in the streets. These are also made of wood and their tops project 2 or 3 feet above the level of the ground. The water is allowed to rise to a certain level in them or to overflow their edges, and the inhabitants procure their supplies by dipping their buckets into them.

In other works which the Japanese have undertaken there may be observed the same want of knowledge of the properties of materials, and the same crude methods of executing work. I have confined myself in this paper entirely to a description of what the people of the country have accomplished without extraneous aid. To what extent foreigners have, in later years, been enabled to educate them, or to develope the building resources of the country, would fitly form the subject of a separate paper, which, if agreeable to the Society, I shall have pleasure in placing before it on some future occasion. But I may be allowed to say here, that while I have felt it impossible to come to any other conclusion than that in constructive art, the Japanese are surprisingly behind us, I do not wish it be understood that I consider this deficiency of knowledge to be due to any want of intelligence on their part. Whatever may have been the causes for the want of attention which has been paid to building, there can be no doubt of the great aptitude and ingenuity of the people, and that, after a few years of well directed education, they will give good proofs of their ability to master all the intricacies of construction as now understood in all civilized countries.

CONSTRUCTIVE ART IN JAPAN
[PAPER 2]

BY R. H. BRUNTON, ESQ.

[Read before the Asiatic Society of Japan on the 13th January, 1875]

IN THE PAPER which I read before the Society on this subject last year, I said that, if agreeable to the Society I would continue the subject on another occasion.

In that paper I gave a description of the evidences of constructive ability displayed by the Japanese before they had availed themselves of the assistance of foreign experts. The continuation of the subject I then thought might suitably consist of a description of the improvements which these have succeeded in effecting. In setting myself to this task, however, I find it is one which is involved in considerable difficulty. In the first place the results which have been attained are so few and of so limited a nature that there is but little to be said concerning them, and in the second place the efficiency or practical advantages of such results are subjects of so debatable a character that, to treat of them from that point of view would form a paper hardly suited to a society of this kind. If therefore, in attempting to fulfil a promise which I formerly made, I have not succeeded in forming a very valuable contribution to the proceedings of the Society, the difficulties surrounding the subject which I have alluded to above are my only excuse.

In the minds of the modern Japanese there seems to be the same desire for the adoption of a dwelling constructed after a European model, as for the adoption of European clothes. They argue, with a shew of reason, that the one is necessary to the other. Thus when sandals or clogs gave way to boots, and the loose flowing robes to the tightly-fitting European dress, it became necessary to discard the old system of squatting on mats and to adopt wooden floors with carpets, and to sit on chairs and at tables. Europeanized dwellings are therefore now common.

The style of building most generally adopted throughout the country in these new houses is a bad copy of the houses to be found in the European settlements. It is almost unnecessary to describe these. They, however, display novel points in the practice of house building which are worth mentioning on that account only. The foundations consist of a stone wall generally about 8 inches thick and 2 feet high. On this wall is laid a wooden sole-plate which is about 6 inches square, and into which the wooden uprights forming the walls of the house are mortised. The uprights, also about 6 inches square, are placed from 2 to 3 feet apart, so that when they are still uncovered they appear like a forest of posts. There are very thin laths placed longitudinally along the uprights at distances of 6 feet or so apart, which are secured to them by wooden pins. Diagonal struts or ties are very seldom used, and the stability of the building is therefore dependent on the stiffness of the different joints in the framework, assisted by the nails used in the different parts of the erection. The roof is formed of timbers very much larger than is required for strength, and is laid with mud and tiles much in the same way as I have described in my former paper as is adopted in Japanese temples. Inside, the

227

houses are generally lined with planks three-eighths of an inch thick, on which wall-paper is placed, the ceilings of the rooms being executed in the same way.

In some of the better class of houses, however, the walls and ceilings are lathed and plastered, but this is by no means general. Outside the walls there are sometimes fixed laths to which square tiles are nailed - the joints of the tiles being pointed with plaster; sometimes the walls are plastered without any tiles, and in those houses which are intended to be of the best description, thin stone flags, of a thickness of about four to eight inches, are built on one another and kept in their places by small iron dogs attached to the woodwork. In some of the houses iron stove-pipes are let through the walls surrounded by a stone, but the more pretentious have fireplaces and chimneys erected with stone in their interiors. These are usually about five or six feet square at the base, are generally badly built, and as they project through the roofs they must be in some cases thirty or forty feet high. They can only be kept upright by the floor or roof beams which project against them, and are a constant source of dread and danger.

This is the new species of building common in Japan, and foreigners are doubtless responsible for it: even at the present day very few houses in the foreign settlements are built after a more secure or substantial style, and in Japanese hands it has, if anything, become worse. When foreigners first arrived in this country they may have had reasons for adopting this method of construction. 1st, It is somewhat similar to the Japanese method, and those who commenced building might have been glad to adopt it on that account, as the work would there be more or less familiar to the only workmen who were available at the time. 2nd, It has the advantage of only requiring the very cheapest and most easily procured materials, and so is well suited for temporary purposes or for hasty erection. 3rd, It is supposed by some persons to be the best construction to resist earthquakes on account of its elasticity and on account of the wooden framework preventing the outside lining of stones or other covering from being precipitated inwards on the occasion of a shock.

The first reason in its favour does not now exist in such strength as formerly, because, although really efficient workmen are still very difficult to procure, there are now in this country many Europeans of experience who, by their superintendence and direction can efficiently cause to be executed almost any species of building. For merely temporary buildings it may still be, on account of its cheapness, the best, but if the construction is to have any pretensions to be a lasting erection, or one which has to afford effectual protection from outside disturbances, I have no hesitation in saying that the system is the most uneconomical. From the fragile nature of the materials which compose the outside casing, whether these are stone flags, tiles or merely plaster, the walls are in want of constant repair, and are never water or air tight. The wooden framework from its insufficient covering decays with great rapidity, and it is in all points excessively and dangerously weak. The third reason in its favour, viz., its efficacy to resist earthquakes, is one which opens out a large field for discussion on which I may have something to say further on.

In copying this system of construction, therefore, I need not say that, in my opinion, the Japanese have been led into an egregious error. And it is really a pity to see such buildings as the new Custom House and the New Town Hall in Yokohama, the new Government offices in Yedo, all of which should be buildings of real stability and durability, built on this principle. These erections have all some pretensions to architecture; they have each cost very large sums of money, and being efforts at improvement in the way of construction, it is most unfortunate that the system adopted was not one formed on a more sound and substantial basis.

Since the great fire which happened in Yedo in 1872, the minds of the local authorities there have been greatly exercised in reference to the construction of buildings which will afford greater resistance to the spread of fire. A very creditable effort has been made in the new Boulevard at Yedo - where small brick houses have taken the place of the slight wooden erections which are general in all Japanese towns. These new buildings are built with brick which has been coated with Portland cement plaster. The walls of the houses are of course perfectly uninflammable, and fire-places with properly constructed chimneys are placed in each wall. These houses present no perfect immunity from fire, but there can be no doubt that from the use of uninflammable material in the walls, and by a well-devised system of construction they offer great checks to the spread of fire, and the danger of taking fire is immensely lessened. The buildings have evidently been designed so as to retain as far as possible the Japanese system of house with open fronts and movable partitions; they are two-storied and contain four small rooms, and have a small verandah in front supported on brick columns. They seem to be a very suitable species of building for the class of Japanese occupying them, and they most certainly present an infinitely better example of building to the people than the European houses which I have just described.

The principal and most important move made by the Japanese Government towards introducing into this country a better appreciation of the art of building and, at the same time, furnishing the country with those results of the ingenuity and labours of our great engineers which have revolutionized the civilized world, is the establishment of the department of public works and the prosecution of the undertakings under its care. The construction of two lines of railway after the English model cannot fail to instil into the minds of the many Japanese employed in connection with them, the advantages of the principles of building adopted in Europe. Although one of these lines does not unfortunately present many features worthy of imitation, the other one in the excellence of the details of the workmanship upon it, whether in brick, stone, or iron, supplies a model of the greatest value. The lesson that these works afford the Japanese should be of the greatest use to them. The various natural products of their country have in them been moulded, formed and brought into combination with each other so as to form structures of precisely the necessary strength and of the most certain durability.

The graving docks and various other works at Yokosuka also present to the Japanese another phase of constructive art from which they may learn the properties and use of another species of material. While the lighthouses, though humble specimens of construction, and labouring under the disadvantage of being placed in such situations that few people see them, afford, I hope, their quota of information.

That the Japanese have not benefited so fully as they might by the lessons given them in the carrying out of such works, I think, can be safely affirmed. This has been occasioned more by a restlessness of mind and want of application than by want of ability. Their natural presumption of knowledge is proverbial, but in addition to this there has not been established to my knowledge any definite system of education among workmen. The methods of manufacture in all countries by means of which the cheapest, the best finished and only reliable articles are produced are well known to consist of keeping each workman confined to one very narrow branch of labour. In this way he becomes expert in that particular line and is able to produce work with a rapidity and of an excellence otherwise unattainable. In building, a stone mason, a bricklayer or a carpenter is obliged to serve a weary apprenticeship of 5 or 6 years, and after that has been completed a long probation of many years on a merely nominal pay before being considered or trusted as an efficient workman. In Japan, on the contrary, bricklayers or masons are procured ready made; a Japanese carpenter is a mason one day and a bricklayer the next. And the introduction of the system of apprenticeship - by which the intelligence and energy of youths are brought to bear on one particular branch of labour - has not, as far as I am aware, been thought of. This defect may be due in a great measure to the exigencies of the country, which has only lately commenced to adopt these improvements, but I fear it is also occasioned by the restlessness of disposition which is a well known feature in the Japanese character.

The materials used for building have been but slightly developed within late years. Wood, which still maintains its supremacy as the principal and the most commonly-used building material, has not improved in quality. It is of a most treacherous character as at present to be purchased in the market, this being due to a want of care in felling the timber, in seasoning it and in drying it. No trials have yet been made so far as I am aware of the strength, durability or weights of Japanese wood, so very little can be added to the information I gave in my former paper concerning it. There can be little doubt, however, that the extra demand for timber caused by the commencement of public works of magnitude, and by a desire for larger and more extensive edifices is causing a denudation of the forests of the country. The rapidity of decay in the material itself and the wholesale destruction to which it is exposed by conflagrations keep up a steady demand for this, the stock building-material of the country; but if to this is added the demand caused by the various improvements which have been instituted, the supply is unable to keep pace with it. With a recklessness which I fear is a characteristic of the people, the forests are being taxed beyond their powers - timber itself has increased in cost within the last five years to twice or thrice its former price - and fears not unnaturally arise of certain climatic changes springing from the clearing of large tracts of country of their former forests.

Bricks were introduced by foreigners some years ago and are rapidly getting into extensive use. They do not, however, as yet attain to that excellence of manufacture which make them at all a desirable building material. The process of brick-making, like all other work requiring skill, can only be carried out to perfection by experts. While the Japanese are well acquainted with the manufacture of various porcelain and terra-cotta articles for use or ornament, and succeed in this most admirably, the making of bricks, so as to turn out both a cheap and reliable article, is a process requiring such entirely different methods of work that their knowledge in the former is not of much avail to them. The selection of the clay, its puddling and the shaping of the bricks are at present all done more or less carelessly and without method, but it is in the burning of the bricks that their principal defect lies. Their kilns are formed in the shape of a cone, and are, when charged, generally filled to the top with bricks. A wooden fire is inserted at the bottom and the heated air is allowed to find its way through the large mass of bricks to the top of the kiln. Those bricks at the bottom, we naturally find, are overburnt and cracked while those on top are quite insufficiently burned. A very small proportion only of really reliable bricks is therefore got from each kiln. I am informed that very excellent bricks are procured in Kobe and Osaka, which are made at Sakai, but I have not had an opportunity of testing these. I have, however, put bricks from various other parts of the country to the ordinary tests, and the results are anything but satisfactory.

The ordinary bricks in use in London absorb after immersion in water not more than one-fifteenth of their own weight, and they withstand pressure of 800 lbs. on the square inch. The ordinary bricks made in Yedo absorb one-fifth of their own weight of water, and will not stand more than 300 lbs. per square inch; that is to say, they are three times as porous as they ought to be and less than half of the strength they should be. Bricks made in various other parts of the country shew almost the same results. The best which I have had opportunity of trying were those made at Hakodate; they stood fully the standard crushing strain but absorbed far too much water. Such bricks are in my opinion quite unfit for building purposes. Their actual strength, when new, may be sufficient for the small erections built in Yokohama, but their porosity renders them liable to rapid disintegration, and to actual rapid destruction from severe frost, while houses built of them must under any circumstances, suffer from continually damp walls.

A very excellent instance of the character of the Japanese bricks may be seen at Hakodate, where a series of large godowns recently erected by the Kaitakushi are in ruins, owing to the splintering and reduction to powder of the brickwork in their walls, caused by one winter's frost. A partial remedy for the badness of the bricks may be to coat the walls with lime or cement plaster, but even with this I fear that sufficient moisture will find its way to them as to have a very deteriorating effect upon them. Another and a very grave reason against the ordinary use of brickwork as at present carried out in this part of Japan is the almost useless character of the lime mortar. Bricks in a building are held together by the lime mortar between them, and if this mortar does not possess the necessary connecting qualities each brick is dependent on itself, and the edifice which is constructed of them is deprived of almost its entire stability. Good mortar requires clean sharp sand and newly burned lime. Such sand in the neighbourhood of Yokohama is difficult to procure and is almost never used, and the Japanese seem to have a pride in using only the oldest lime, which from long keeping has entirely lost its virtues.

The system of construction best suited to withstand earthquakes is a consideration which should always hold a prominent place in the design of any erection in Japan. I have not been able, though I have made considerable efforts to do so, to procure any information either to verify the particulars I gave in my former paper regarding earthquakes or to make additions to that. While we are perfectly aware, therefore, of the liability of the country to shocks of destructive violence, we are not aware of the nature of the shocks or the localities in which they may be expected to be most severe. It is to be hoped, however, that with the assistance which the Japanese Government now possesses, both as regards instruments and professional men, that before long we may have a regular system of observations affording us this information.

I have formed the opinion that the heavy roof and the light framework in Japanese erections are ill-suited to withstand these shocks, and I believe my opinion to be sustained by the truest principles of mechanics. This, however, is hardly the place to enter into a disquisition upon that subject. I am also of opinion that a solid erection, properly constructed, will afford the greatest safety during an earthquake and at the same time is the only one which will give reasonable security against fire, wind or the other natural disturbances. An appliance has been devised by a well known English engineer for the purpose of counteracting the disturbing force of an earthquake, the principle of which is very simple. It was said by him that the movement given to the foundations of a building is transmitted with accelerating force to its summit, and that to destroy this the simplest method was to make a break in the continuity of the structure. The designer, therefore, proposed that buildings should be made in two parts, the lower part to be firmly embedded in the earth, the upper to rest on balls which are made to roll in inverted cups. A sudden movement in the lower part would not, then, be transmitted to the upper on account of this break or joint in the structure, and the experiments made shew that in point of fact this theory was perfectly correct. He procured this idea from seeing in Japanese drawings the uprights of their houses resting on round stones, imagining this to be done in order to give them as slight a hold of the earth as possible. But from enquiries I have made this does not seem to have entered the minds of the Japanese, and the only way they can account for the uprights being placed on round stones is to keep the wood away from the moisture of the ground and because round boulders are more easily and cheaply procured than square stones.

The impracticability of this scheme, which however deserves a fuller trial than it has yet had, arises, in my mind, from the fact that a house resting on balls is liable to be swung and rocked about in gales of wind to such an extent as to render it unfit to live in. I may mention that the tables on which the apparatus of some of the Japanese lighthouses rest are constructed with this joint, but they have been found to be unsuitable for the reason I have just stated, viz., that when touched or trod upon they shake and roll too much. It is quite possible, however, that some alterations might be effected on the design to obviate this difficulty, and if this were accomplished, there can be no doubt that it would afford great immunity from earthquake shocks. Any scheme which will afford this should be welcomed in Japan, but at the same time it is in my opinion a great mistake to sacrifice the whole comfort and the safety of dwellings from their dread enemies fire and wind to a supposed protection from so remote a contingency as a severe earthquake is in Japan. The building most truly suitable is one which to the fullest extent provides a protection from both these disturbances.

Wooden houses, on the supposition that they are best during earthquakes, are not well fitted to withstand the other disturbances to which they are much more frequently exposed - and solidly constructed houses, while they are less inflammable, and less liable to damage from bad weather, whether they are well calculated to resist an earthquake shock or not, at all events have this advantage, that they will always afford their inmates time to escape. The more solidly constructed a building is the longer it will remain standing, even though solidity is no perfect protection against the rage of an earthquake.

The matter of cost is an important one in regard to works of this kind, and there is not an unreasonable fear that to erect a solid building means a large expenditure. This, under present circumstances in Japan, is probably the case, though not to that extent which might be imagined. The present want of good material and the dearth of efficient workmen enhance the cost of good work, but we must hope that these hindrances will soon be removed, and that in a short space the towns of Japan will consist of something better than rows of tinder-boxes.

THE

JAPAN LIGHTS

BY

RICHARD HENRY BRUNTON, M.INST.C.E.

WITH AN ABSTRACT OF THE DISCUSSION UPON THE PAPER

EDITED BY

JAMES FORREST, ASSOC.INST.C.E.

SECRETARY

By permission of the Council.
Excerpt Minutes of Proceedings of The Institution of Civil Engineers.
Vol. xlvii, Session 1876-77 - Part i.

ADVERTISEMENT

THE INSTITUTION OF CIVIL ENGINEERS

Sect. I - MINUTES OF PROCEEDINGS

November 14, 1876

GEORGE ROBERT STEPHENSON, President
in the Chair

No. 1,451 - 'The Japan Lights'. BY RICHARD HENRY BRUNTON
M.Inst.C.E.

THE JAPAN LIGHTS

SKETCH SHOWING THEIR POSITION, AUGUST 1875

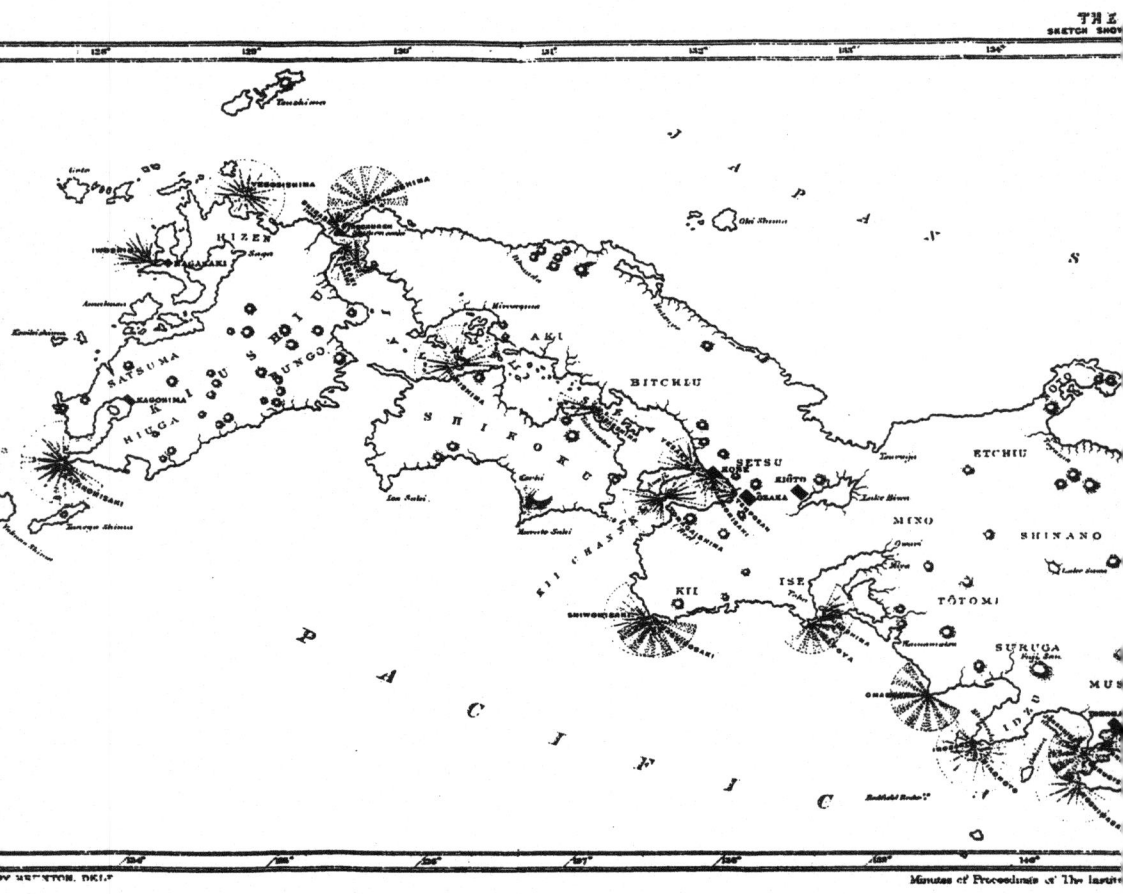

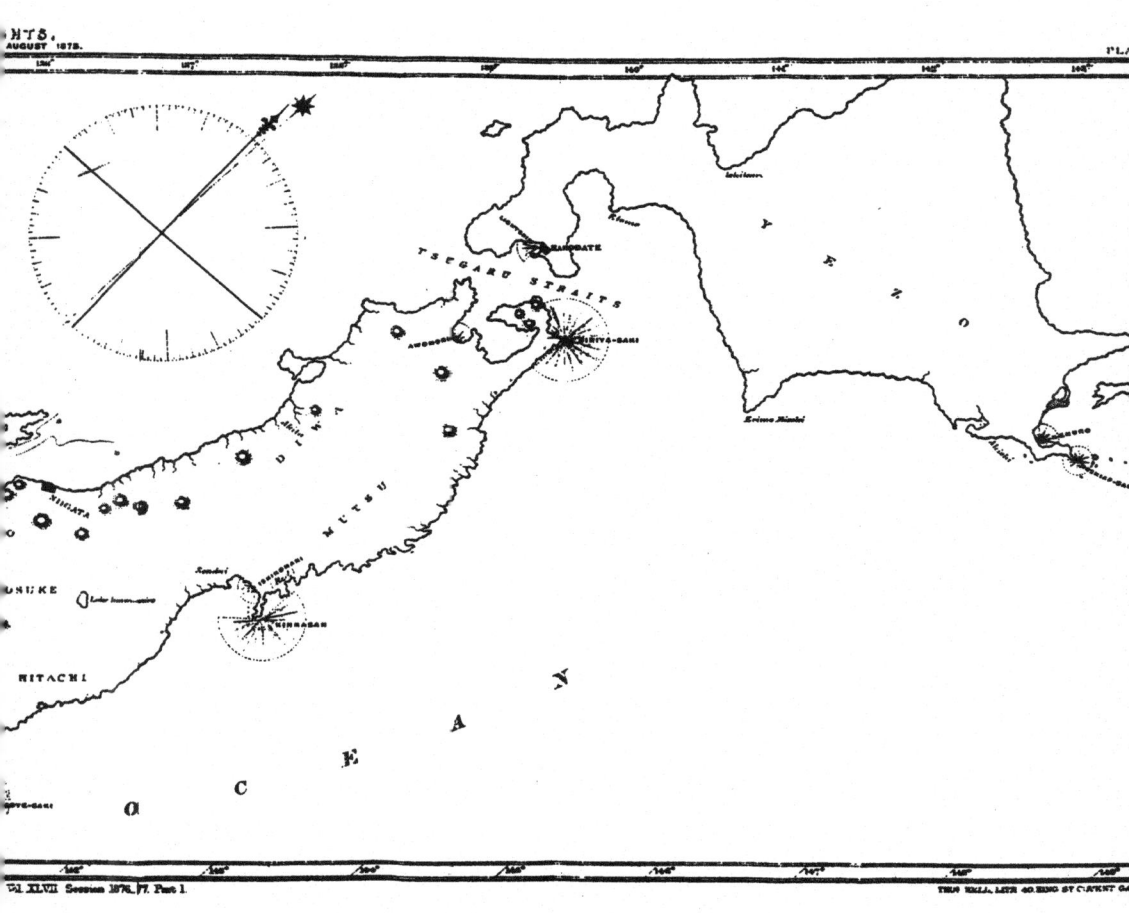

235

THE

JAPAN LIGHTS

BY

RICHARD HENRY BRUNTON, M.INST.C.E.

THE Government of Japan, in entering into friendly relations with foreign powers, stipulated that certain ports were to be open, and that mercantile operations were to be confined to those places. These ports are:
1 Hakodate, in the island of Yezo. [sic]*
2 Yokohama, 20 miles from the northern capital of the country, Yedo, or Tokio.
3 Kobe, 50 miles from the southern capital, Kioto, and near the entrance to the Inland Sea.
4 Ozaka, between Kobe and Kioto, being 20 miles from the former.
5 Nagasaki, in the south-west portion of the island of Kiusiu.
6 Niigata, on the west coast of the Main Island.

The principal trade is at Yokohama, where from two hundred to three hundred foreign-owned vessels of large size annually resort. The next places in importance are Kobe and Nagasaki, at each of which about one hundred and seventy vessels arrive every year. Hakodate is in a high latitude, and its trade is almost entirely confined to the summer months, but it is at no time of any great extent. Niigata and Ozaka are visited by few ships. All these ports, however, are the rendezvous for a considerable native trade, which is every year increasing.

The foreign powers arranged, and it forms one of the clauses in the existing treaties, that the Japanese Government should 'provide the treaty ports with such lights as may be necessary to render secure the navigation of the approaches to the said ports'. On the 17th of November 1866, Sir Harry S. Parkes, K C.B., H.B.M. representative, directed the attention of the Japanese Government to the existence of this stipulation, and requested that steps might be taken to carry out its purport.

The French and American ministers shortly afterwards made similar representations, and a favourable reply was received from the Japanese ministers early in December of the same year. Sir H. S. Parkes had previously received suggestions from naval officers and other relative to the most important sites for the lights, and these he laid before the Japanese Government. The erection of eleven lights was agreed to, and Sir H. S. Parkes was requested to use his influence with Her Majesty's Government, in order that the necessary apparatus and the assistance required to establish lighthouses might be obtained. The matter having been referred to the Board of Trade, that department consulted the Trinity House, and eventually, on the 6th of November 1867, Messrs. D. and T. Stevenson, MM.Inst.C.E., the Engineers to the Commissioners of Northern Lights, were desired to select suitable persons to undertake the design and construction of the lighthouses, and to introduce the lighthouse service into Japan. The Board of Trade had the general management and supervision of the arrangements, while the Messrs. Stevenson, who pointed out the difficulties likely to be experienced from the constant recurrence of earthquake shocks in Japan, had charge of the construction of the apparatus for six of the lights which had been ordered.

The Author received the appointment of Chief Engineer in February 1868, and he arrived in Japan in August of the same year.

Out of more than twenty-five sites suggested by those to whom the matter had been referred in Japan, five points appeared to be generally concurred in (Plate 1). These were:- A second order revolving light at Oshima (Kashinosaki); a first order fixed light at Shiwomisaki; a first order fixed light at Iwoshima; a first order fixed light at Satanomisaki; and a lightship at Hakodate.

It was found, however, impossible to select from the various suggestions the most advantageous positions for the remaining six lights authorized by the Japanese Government. A proposal was therefore made by Sir H. S. Parkes to complete the necessary information by a mixed Commission of the senior naval officers of England, France, and America then in Japan, the Japanese Government being also represented on it. This Commission consisted of Captain Hewett, of the English navy; Commodore Goldsborough, of the United States navy; and Commandant Amet, of the French navy. The Japanese war-steamer 'Fujiyama' was placed at the disposal of these officers, and they visited the various points in November 1867. They agreed in recommending a first order fixed light on Noshima; a first order fixed light on Mikomoto (Rock Island); a second order flashing light on Tsurugisaki (Sagami); a third order fixed light on Kannonsaki; a red floating light at Yokohama harbour; as well as several buoys and beacons in the Gulf of Yedo.

Some time previous to the formation of this Commission the apparatus for several lights had been ordered from France through a staff of French engineers, who were then engaged in forming an arsenal at Yokoska, on the west coast of the Gulf of Yedo, about 10 miles from Yokohama. These were intended to light the approaches to the arsenal; but the Commission recommended that they should be placed on those points most urgently requiring illumination. It was afterwards found that the only spots suitable for them in the Gulf of Yedo were Noshima and Kannonsaki. The first order fixed apparatus from France was accordingly placed at Noshima, and the third order fixed apparatus at Kannonsaki, and these two lighthouses were erected under the superintendence of the French engineers above mentioned.

On arriving in Japan, the Author examined the sites of the lighthouses which had been decided on; and, as they were spread over 1,500 miles of coast, and there were no means of internal communication, Admiral Sir Henry Keppel, then on the Japan station, was induced to grant the use of H.M.S. 'Manilla' for a voyage of inspection. In addition to collecting information and taking the requisite observations in regard to those points already decided on, the Author received instructions to inspect and report upon the best method of lighting the Inland Sea and the approaches to Kobe and Ozaka; for which purpose the Japanese Government, in the beginning of 1867, had authorized the apparatus for five additional lights. The Inland Sea, separating the main island from Kiusiu and Shikoku, is about 250 miles long, and, at some places, it is 50 miles wide; it is filled with several thousand small islands, and navigation is carried on through recognised channels between them. It may be judged, therefore, that to place a light on every point in the Inland Sea requisite to render navigation through it practicable, on dark nights, would be a work of magnitude and of doubtful utility. On the other hand a few lights might be so placed that, with their assistance, a considerable portion of the sea would be navigable at night, and vessels would be guided to places where they might anchor with safety and wait for daylight.

The proposals made by the Author were based on the following principles. 1st. To put no light where the headlands are bold and well defined, or where no hidden dangers exist. 2nd. Where there are difficult and dangerous channels, through which vessels could not proceed in dark weather, to place lights to lead them into a safe anchorage where they can wait for daylight; and, if possible, in such a position that advantage may be taken of them for going through the channel. 3rd. Where a light would render a channel easy to pass through, which, without it, would be difficult, or where a light is likely to be required to give mariners a clue to their position, regarding which they may, at night, get easily confused among the numerous islands, a light should be placed.

Having regard to these considerations, the Author fixed upon the following points at the approaches to Kobe, and Ozaka, and in the Inland Sea, as being the most advantageous:- A third order fixed light at Tomagaishima; a fourth order fixed light at Temposan (Ozaka); a fourth order fixed light (red) at Wadanomisaki (Kobe); a first order fixed light at Yesaki (Awaji); a fourth order fixed light at Nabeshima; a third order fixed light at Tsurishima; a third order fixed light at Hesaki; and a fourth order fixed light at Rockuren. These proposals were laid before various nautical authorities. They received general approval, and the erection of the lighthouses was ultimately authorized.

Soon after the completion of the two or three lighthouses, the Government showed its appreciation of them by deciding that the old system of wood fires should be abolished, and that properly illuminated lighthouses should take their place. A notification was issued by the Council of State calling upon the local authorities to name the places of which local lights were needed, and instructing them that the old system of wood fires must be discontinued. The result of this notification is that many applications for small lights have been received from all parts of the country. But as the Imperial Government has, except in a few important cases, refused to bear the expense of these lights, and as the local governments are generally unable to do so, not many of the proposals have as yet been carried out. Eleven such lights are now established, and from ten to fifteen others are under consideration.

Those which have been completed are:- A staff light on the pier in Yokohama harbour; a fifth order fixed light at Shinagawa, in Yedo harbour;[1] a fifth order fixed light at Jokashima, near the entrance to the Gulf of Yedo;[1] a sixth order fixed red light at Irosaki, near Mikomoto (Rock Island); a fourth order fixed light at Sugashima, the entrance to Toba harbour; a fourth order revolving light at the entrance to Matoya harbour; a fourth order red fixed light at Shirasu, west of Shimonoseki Straits; a staff light on Noshiaf, the easternmost point of the island of Yezo; a staff light in Nemuro harbour, in Yezo; a staff light at Awomori harbour; and a staff light at Ishi-no-maki, at the mouth of the river Kita-kami.

The Government, desirous of further perfecting the illumination of the coast, decided on erecting several ocean lights on those points most in need of illumination. These are in addition to the lights already specified, which were demanded by the treaty powers. A full and complete scheme for the illumination of the entire coast was drawn up, and those sites which were considered most important were first dealt with. Seven of these lighthouses are completed, and the positions for thirty others have been visited, examined, and reported on. Before finally deciding to erect any lighthouse, the Author submitted his proposals to the naval authorities in Japan at the time, and to many captains of vessels trading on the coast. Captain St. John, H.M. surveying ship *Sylvia*, gave especial attention to the matter, and rendered great assistance by his advice.

These seven lights are situated as follows:- A second order fixed light at Siriyasaki, the north-easternmost point of the main island; a first order revolving light at Kinkasan, east coast, north of Yokohama; a first order revolving light at Inuboyesaki, east coast, north of Yokohama; a first order revolving light at Omaesaki, south coast; a first order flashing light at Kadoshima, west coast, north of Shimonoseki; a second order fixed light at Yebosishima, west coast, south of Shimonoseki; and a fourth order fixed light at Haneda, Gulf of Yedo.

The apparatus for these lights, as well as for the harbour lights already mentioned, was ordered, without the intervention of the Board of Trade, through Messrs. D. and T. Stevenson, who, in September 1871, were appointed Consulting Engineers to the Japanese Lighthouse Department.

A complete list of the lighthouses, lightships, buoys and beacons established by the Author on the coast of Japan will be found in the Appendix.

In designing the lighthouses the most important point for consideration was the liability of the country to the periodical recurrence of earthquakes. No definite records had been kept of these disturbances, nor could data be procured as to the localities most subject to them. It is well known, however, that slight shocks are of frequent occurrence, and that, during the present century, fifteen shocks of a destructive character have been experienced. These were felt in different parts of the country, the northern capital, Yedo, being most subject to them, where five out of the fifteen occurred, the remaining ten being pretty equally distributed. The palace of the Emperor at Kioto was destroyed by an earthquake in 1828, as well as most of the temples with which that city abounded. In 1846, in the province of Shinano, on the west coast, the earth swallowed up eighteen houses, five thousand dwellings being at the same time destroyed. The most recent destructive earthquake, in the year 1855, was chiefly felt at Yedo, when the trembling of the earth continued for one month, and eighty severe shocks occurred; about one hundred and twenty thousand lives were supposed to have been lost on this occasion.

Messrs. D. and T. Stevenson, on whom the duty of designing the apparatus for the lights devolved, deemed it necessary to 'take some steps to secure these from derangement, and the lamps from partial or total extinction on the occurrence of the many modified shocks which visit the country'. The arrangements carried out have been thus described by Mr. David Stevenson:- 'It is evident that any sudden lateral motion of the earth, on which a building rests, must be communicated to the foundation of the structure, and thence through all the rigid and unyielding materials of which it is composed to its very summit, where the violence of the shock will be aggravated by the greater elevation of the highest point of the building above the source of motion. On fully considering this action of earthquakes, it seemed to me that what was required to neutralise their shocks was a *break* in the continuity of the rigid parts forming the structure, so as to prevent the propagation of the shock.... The plan I proposed for this purpose, which may, for brevity, be termed an *aseismatic joint*, is the introduction of spherical balls of bell-metal, working in cups of the same material, placed between two platforms, the lower cups being fixed to beams forming the foundation, and the upper cups being fixed to the lower beams of the superstructure, thus admitting, within a limited range, free motion of the upper over the lower part of the building.'[2] The Board of Trade sanctioned this arrangement for the lights first ordered, the apparatus for these being now placed on tables having an aseismatic joint on the principle described (Plate 2, Fig. 1).

The Author's experience of the contrivance is not altogether favourable.

He finds that, while the free motion of the upper over the lower part of a structure may neutralise the effects of an earthquake shock, such free motion will, at other times, occasion inconvenient results. Thus, were such a joint placed above the foundations of a lighthouse tower, the pressure exerted by a gale of wind on the superstructure would give rise to a motion probably equally as distressing as a severe earthquake. In the same way a person stepping on one of the aseismatic tables, for the purpose of trimmming or cleaning the lamps, causes the upper part to roll to such an extent that the lamps become deranged, and in the case of revolving lights the regular motion of the clock-work machinery is destroyed.

Messrs. Stevenson introduced a stout spring in a vertical position, the lower part of which was securely fixed, while the upper part was attached to a ball working in a socket in the centre of the upper table. This was intended to regulate the movement, but it was only partially successful. The problem is to form a joint sufficiently sensitive to move freely on the occasion of an earthquake, but so stiff that no inconvenient motion will be occasioned by ordinary disturbances. Messrs. Stevenson's design was, for the reasons named, not adopted in any of the other lighthouses.

From the observations of those who have examined the effects of earthquakes, and have given their attention to the subject, it may be accepted that there is a double motion in all shocks, the first movement tending to overthrow, while the second tends to restore to equilibrium. If, therefore, a building could be constructed with sufficient power of resilience, its overthrow or destruction could not occur so long as the oscillation caused by the earthquake did not go beyond its limit of equilibrium. This seems to have been the leading principle in the native architecture of the country. The houses are constructed with a light wooden framework, without diagonal struts or ties of any kind. The roofs are heavy, and the uprights which support them are mortised into horizontal beams at the top and at the bottom. The buildings are evidently designed to have a maximum of elasticity, and they can, no doubt, be, and often are, moved off the vertical to a great extent without fracture. Their efficacy in resisting severe earthquakes, however, is questionable. The system further seemed open to the objection of excessive weakness, and consequent inability to withstand the other disturbing forces to which a building is exposed. The only alternative method that seemed feasible was to give the lighthouses great weight and solidity, thereby adding to their inertia and checking their oscillation. It has been laid down by Professor Palmieri of Naples, probably the most experienced earthquake observer in the world, that although solidity and strength in a building do not afford perfect protection against an earthquake, still so long as fracture does not occur, absolute overthrow is almost impossible. This principle of solidity combined with strength was the one adopted by the Author in designing the lighthouses, and in carrying out the different details of their construction.

The various lighthouses were constructed of stone or brick, wood or iron, as found most suitable for the locality. They are generally circular at the base, and the walls have a straight batter on the outside, and are plumb on the inside. Forming a semicircle round the bases are two store-rooms, one for oil, and the other for dry stores. Paint, fuel, and additional store-rooms are erected in the grounds. The lightkeepers' dwellings are of stone or brick, and contain from six to eight rooms: the kitchens and outhouses are in separate buildings. The grounds are surrounded by a stout fence or wall, about 8 feet high.

The stone lighthouses have all been constructed of ashlar masonry. Most of the stone employed was granite, of excellent quality. In some cases a volcanic clay-stone, of a greyish appearance, and of a tough hard character, has been used; but there are evidences of slight deterioration in this stone on the exposed surfaces. The courses have, in all cases, been 12 inches thick; and headers, from 4 feet to 2 feet wide, have been put through the walls at every third or fifth course. The stretchers are also of large size, there being, in most cases, not more than two in the thickness of the walls.

The brick lighthouses are amongst the latest erected. The making of bricks in Japan has only been recently introduced, and their manufacture has not yet obtained such perfection as to warrant their adoption where any other material is procurable. All the bricks were specially made by men employed by the Department for the purpose. The bricks were generally well formed, and of good reliable clay; but the native workmen were negligent in carrying out the process of burning, and the bricks suffered in consequence. Every precaution was taken to reject faulty ones. Those which were used resisted a pressure of 700 lbs. on the square inch. Their porosity, however, was considerably greater than was desirable, as they absorbed about 10 to 12 per cent of their weight of water. In order to render their liability to deterioration as slight as possible, the outsides of all brick towers have been coated with Portland-cement plaster. 'Flemish' bond was adopted, and hoop-iron bond was inserted between every fifth course. The lintels, soles, and rabbets of the doors and windows are all of stone. For about 13 inches from the face of the walls the courses are jointed with Portland-cement mortar, the remainder being laid in lime mortar.

Special precautions were taken regarding the lime supplied for the works. The Japanese were formerly ignorant of the use or properties of lime mortar; and although limestone was burned by them for purposes of their own, their custom was to keep it, after it had been slaked, in straw bags for indefinite periods, and to sell it in this state. They had never adopted the system of mixing it with sand and using it as a bed for stones, the principal use to which it was put being as plaster, when it was mixed with boiled seaweed. At most of the lighthouse works lime-kilns were erected, and the limestone burned on the spot. The lime was generally pure, and good mortar was formed by mixing 1 part of it with from 3 to 5 parts of sand.

The wood adopted for the main beams, uprights, &c., of those lighthouses which were built of timber is a native production named 'keaki' (*Planera Japonica*). It is hard, only a little lighter than water, with a close grain, and when of good quality, of an endurable nature. But it is difficult to procure good keaki, owing to the absence of restrictions in the felling of timber. Much of the wood offered for sale is unripe or full of sap, and little is to be found which has undergone seasoning, to which process but small attention is paid by the natives. Keaki is the only hard wood grown in the country in sufficient quantity to be available for building. The favourite soft wood, for floors, linings, doors, and windows, &c., is called 'shinoki' (*Chamoecyparis obtusa*). It has a beautiful white grain, and is very lasting. The other woods employed in the lighthouses are 'sugi', a species of cedar (*Cryptomeria Japonica*), and 'matsu', a kind of pine (*Pinus densiflora*). These are considered inferior, and decay quickly when exposed to the weather. They, however, are durable when kept dry, and have been used for the trusses of roofs and for floor-joists, &c. Most of the wooden lighthouses are octagonal at the base, the eight uprights, at the angles, resting upon square blocks of stone, founded in the earth. There is also a centre upright, from which horizontal beams radiate towards the eight outer uprights. Between these latter there are diagonal beams, crossing each other, and horizontal beams at the top of the diagonals. The beams are all secured by iron straps and bolts. The drawings for these towers were made after designs furnished by Messrs. D. and T. Stevenson. The scantlings of the timbers are greater than would have been necessary with the ordinary hard woods used in England, although it is generally believed that keaki would compare favourably with any of them in point of strength.

All iron used in Japan up to the present time has been imported from Europe, and structures formed of it are necessarily costly. Three lighthouses only have, on this account, been made of iron, and then it was adopted on account of the inaccessibility of their sites. It was found that the different parts of the lighthouse, when of iron, were more easily handled, and, after they had been fitted in their places, were more rapidly put together, than if any other material was employed. These iron towers are from 20 to 30 feet high, and are entirely cased with $\frac{1}{4}$ and $\frac{3}{16}$-inch plate, riveted to the beams. They are constructed with eight or more girder uprights, consisting of two angle-irons, 3 inches by 3 inches by $\frac{1}{2}$ inch, and a web and flange of $\frac{3}{8}$-inch plate, 12 inches wide. These support the girder beams of the different floors, as well as the lantern and apparatus. Small uprights at the angles of the towers rest upon a wrought-iron sole-plate bedded on a stone foundation. One iron screw pile lighthouse, erected at Haneda, was designed by Messrs. Stevenson, constructed in England, and sent to Japan in pieces.

There were unusual difficulties in the construction of some of the lighthouses. For instance, Mikomoto, or Rock Island Lighthouse (Plate 2, Fig. 2), has been erected on an isolated rock about 2,000 feet long by 450 feet wide, by 100 feet high, at its summit. It is situated off the harbour of Shimoda, about 80 miles south-west of Yokohama, and at a distance of 6 miles from the nearest land. All vessels on their way to Yokohama from the south pass this rock; and, as there are rocks between it and the shore, it is an extremely dangerous locality, and an important position for a light. The sides of the island are precipitious, and in heavy gales of wind green seas dash right over it. It is composed of a hard, brittle, igneous rock, worn to sharp points by the weather. The currents, which run with great velocity in the neighbourhood, cause a constant turbulence in the sea, and it is, further, subject to frequent gales. The locality is also believed to have suffered severely from earthquake shocks. From the exposed position of the rock, the construction of this lighthouse was a work of difficulty. At Shimoda a hard clay-stone was found, and a quarry was opened for the works. Limestone was discovered in close proximity, and a kiln was erected on the rock, where the stone was burned. It yielded an excellent lime. Communication with the rock was carried on by native-built boats, worked by the natives with extraordinary skill. The lighthouse is of stone, 58 feet high to the sole-plate of the lantern. It is in the shape of a truncated cone, and is surmounted by a capital having twenty-four Gothic arched recesses round it. The diameter at the base is 22 feet and at the top 16 feet. The thickness of the walls at the base is 6 feet and at the top 3 feet. It is fitted with a spiral staircase of keaki. The light shows all round the horizon, and a red ray of 55° is inserted, which covers all dangers between it and the shore. The work of cutting away the rock to prepare for the foundations of the tower was commenced in April 1869, and the lighthouse was first illuminated on the 1st of January 1871.

Yebosishima (Plate 2, Figs. 7 and 8) is a conical rock 60 miles to the south-westward of Shimonoseki Straits and 10 miles off the mainland. It is passed by all vessels proceeding between the Inland Sea and Nagasaki. The island is 120 feet high, and is composed of hard basalt. Its sides are precipitous.

Immense boulders, weighing upwards of 50 tons each, were found on its summit, and these had to be removed before sufficient space could be obtained for the buildings. Owing to the inaccessibility of the site an iron tower was erected. It is octagonal, 32 feet 6 inches high, and the extreme width is 23 feet 6 inches at the bottom, and 18 feet 6 inches at the top. A stone dwelling-house and a water and paint store were erected on the rock. Winding steps were cut in the face of the rock, and a tramway was laid by which the materials were conveyed to the summit. The materials were brought, as at Rock Island, in native boats, and were lifted out by shear legs, made to project over the boats, as these could only lie at some distance off. No landing could be effected unless the water was quite smooth, and progress was on this account much delayed. The works were commenced in August 1873, and the light was first exhibited on the 1st of August 1875. A relief station has been established at the nearest point of the mainland, on one of the headlands at the entrance to Yobuko harbour. Two lightkeepers alternately live here, leaving three on watch at Yebosi, and a system of signals is arranged between the relief station and the lighthouse.

Satanomisaki (Chichakoff), the most southerly point of Japan, is the headland which vessels coming from the China seas generally make first. The headland is lofty and precipitous, and is covered with thick brushwood; but no site could be found on it for a lighthouse, to serve all the intended purposes. Directly off the point, however, there are two small islands, the outside one of which is about 300 yards distant from the mainland and 180 feet high. The point is exposed to the full strength of the Japan stream, which not unfrequently attains a speed of 3 or 4 miles an hour. This causes a constant disturbance in the sea, and between the outside island and the mainland there is generally a boiling surge. The island is pinnacle-shaped, and its summit had to be lowered for nearly 40 feet, before sufficient space could be obtained for the lighthouse tower. The dwellings for the lightkeepers were erected on the mainland. Owing to the disturbance in the sea caused by the currents, communication between the dwellings and the lighthouse was difficult. A wire rope was originally stretched across, first to the middle island and thence to the lighthouse, along which a cage was propelled by friction pulleys worked by a handle in the cage. This was used as a means of transport for men and materials for several years, but owing to a disinclination of the lightkeepers to use it, caused by nervousness from being suspended at so great a height, it was eventually abandoned. Communication is now carried on by boats, when the water is sufficiently smooth to enable a landing to be made. The tower is of iron, and is 17 feet 9 inches high to the sole-plate of the lantern. It is hexagonal, 24 feet in extreme width at its base and 16 feet at its summit. The lower floor of the tower is set apart as a sleeping and living room for the lightkeepers stationed on the rock, and space is also provided to enable them to keep sufficient provisions to last until relief arrives from the mainland. The other floor of the tower is used for the purposes of the light.

The highest lighthouses are the brick lighthouse at Noshima, 85 feet high to the sole-plate of the lantern; the brick lighthouse at Inuboye, 80 feet high to the sole-plate of the iron parapet; the brick lighthouse at Siriyasaki, 82 feet 6 inches high to the sole-plate of the iron parapet; the wooden lighthouse at Shiwomisaki, 61 feet high to the sole of the lantern; and the stone lighthouse at Kadoshima (Plate 2, Fig. 3), 74 feet 7 inches high to the sole of the iron parapet.

The lightships were built, under the superintendence of the Author, by native workmen. They are 70 feet 8 inches long between the perpendiculars,

18 feet 10 inches breadth of beam, 9 feet 2 inches deep from the deck to the top of the limbers, and are 130 tons burthen by builders' measurement. They are fitted with two decks, having a clear headway of 6 feet 6 inches between them. The lower one, running the whole length of the ship, is used as a berth-deck for the accommodation of the captain and crew. Below are the water-tanks, cable-lockers, ballast space, fuel and lumber stores. The keel, stern-post, stem, timbers, &c., are of keaki. The outside planking is also of keaki, 2 inches thick. The upper deck is of shinoki 2½ inches thick; and the lower deck of sugi, or cedar, 2 inches thick. The vessels are fastened with ¾-inch copper bolts, and are sheathed with Muntz metal, weighing 26 oz. to the superficial foot, up to 2 feet above the water-line. They have two masts. The lantern, containing the light apparatus, is hoisted on the mainmast, which is 16 inches in diameter, and 40 feet high. They are flat on the floors, and are made with a bow specially adapted for riding in a heavy sea after the model of the Trinity light-vessels. They are moored with two 30-cwt. anchors, having 60 fathoms of 1½-inch standard cable on each.

The beacons, built on rocks only awash at low water, are constructed of solid ashlar masonry, the stones in each course being made to radiate from the centre. They are 8 feet in diameter at the base, and 4 feet at the summit; the height being 20 feet. In the first six courses the stones are secured by hard wood joggles; and the coping stones, or the balls which surmount the structure, are secured by a 2-inch bolt, which goes through 7 feet of the masonry.

The buoys were constructed of iron, likewise from the designs of the Author, by native workmen. They are of the ordinary shapes of nun, can, and hollow-bottomed buoys. Their moorings consist of two lengths, of from 15 to 30 fathoms, of ¾-inch cable, attached to two 5-cwt. anchors, one being placed up, and the other down stream. The anchors made by Japanese workmen are very efficient.

The glass-work of the apparatus for the different lights has been manufactured from the designs of Messrs. D. & T. Stevenson, by Messrs. Chance Brothers, Birmingham; Messrs. Sautter, Lemonier, and Co., and Messrs. Barbier and Fenestre, Paris. The lanterns, machines, reflectors, reflector-frames, &c., have been principally executed by Messrs. Milne and Son, and Messrs. Dove and Co. of Edinburgh.

Besides the aseismatic tables in the first lighthouses, the Messrs. Stevenson were induced to adopt a species of apparatus of as unfragile a nature as possible. They had fears that the sudden shocks, to which the lighthouses would be exposed, might derange the delicate glass-work in a dioptric apparatus, and they consequently decided to use metallic reflectors. These are arranged on frames secured to the aseismatic tables, and they were supplied for all the lights included in the original order. In addition to this the lanterns were made as low as possible, and of an increased strength. The daylight height of these is 6 feet only; whereas the ordinary height of first order lanterns is 9 feet 9 inches.

While there can be little doubt that these measures tend to lessen the risk of fracture or derangement from earthquakes; and while the use of reflectors offers this additional advantage, that in case of the partial destruction of the apparatus, the light can be at once reinstated by keeping in stock a few spare reflectors, lamps, &c., other considerations render the adoption of these a matter of doubtful expediency. The whole country is subject to severe earthquakes; but that any single lighthouse should experience a shock of such severity as would derange its optical apparatus, is a remote contingency, and not likely to occur more than once in a century. It may further be assumed, that such a shock as would fracture an apparatus would, in all probability, prove destructive to the lighthouse tower in which it was placed. After the occurrence of such an earthquake, therefore, the permanent light must necessarily be extinguished until steps were taken to replace both the tower and the apparatus, and the accomplishment of this would, under any circumstances, occupy some time.

The Author accordingly thought it unnecessary to adopt extraordinary precautions in regard to the lantern or apparats, when he felt it to be impracticable to use these in the construction of the towers; and the adoption of a reflector apparatus, and a lantern of small dimensions, offers many disadvantages.

1 The cost of a reflector apparatus, such as was sent to Japan, is somewhat more than that of a dioptric apparatus.

2 The polishing of reflectors requires greater skill and attention than the cleaning of glass, and the liability to damage during the process is greater. This is an important consideration in a country where the lightkeepers are only partially trained, and of not very attentive habits.

3 The consumption of oil in reflector lights is about 25 per cent higher than in dioptric lights in relation to the amount of light produced; and the consumption of wicks and lamp-chimneys is also greater.

4 The excessive heat given out by the flame in reflector lights impairs the ventilation of the lantern. In the first order lights, twenty-eight reflectors are used to transmit an equal light to every part of the horizon; and as in each there is an Argand burner, the accumulated heat from these renders a proper ventilation of the lantern a difficult matter. This is specially noticeable on fine summer nights.

5 The lowness of the lantern confines the air-space, and tends to increase the heat and to check ventilation.

The Author therefore concluded that, while the risk of adopting dioptric apparatus was doubtful, its advantages over reflector lights were important; and he recommended that, for future lights, the ordinary dioptric apparatus, and the size of lantern in use in Europe, should be adopted. His recommendations on this point were finally approved and acted upon.

All the lanterns have been formed with diagonal astragals, similar to those adopted in the Northern Lights. They have been fitted with a cast-iron cornice-plate and sole-plate, to which the astragals, made of gun-metal, are screwed. The roof is dome-shaped and is double; the inner casing being made of No. 16 B.W.G. sheet copper and the outer of No. 14 B.W.G. sheet copper. The ventilator on the top of the dome is of the same design as that used on the Northern Lights, having a couple of funnels each covered by caps, surmounted by a hemisphere of copper, through which arrangement the heated air finds exit. Storm-panes are supplied to each lantern, and are provided with clip screws by which they can be attached to the lantern astragals, so that on the breakage of a pane they can be screwed in their places without delay.

Stevenson's holophotalized reflectors adopted in the catoptric lights, are parabolic, 21 inches in diameter at their mouths, with a spherical mirror at the back, and a lens in front of the flames. The glass-work consists of a central lens and four totally reflecting zones. These are made so that no lights from the flame shall escape without being acted upon. The frames on which the reflectors are grouped have been constructed to carry two tiers of reflectors; those lights which show all round the horizon have fourteen on each tier. In fixed lights they are so arranged, that the axes of the apparatus on the upper tier fall midway between those of the lower tier; in this way insuring an equable light in every azimuth. In revolving lights the frames are made with from six to eight faces, and two or more reflectors are placed on each face. The frames are of wrought iron, and are fitted with brackets on which the reflectors rest, and which allow of the fourteen reflectors in fixed lights being placed round the circle.

The lightship lanterns surround the mainmast of the vessel. They are octagonal in shape, and are 5 feet in width from glass to glass, the daylight height being 2 feet 2½ inches. Ventilation is procured by eight tubes, on the roof, 3 inches in diameter, which are surmounted by a hemispherical cowl. There are also in the cornice-plate seventy-two holes each 1 inch in diameter, and over these are sliding plates, with corresponding holes, while there are eight ventilators in the floor of the lantern. There are eight sets of apparatus in the lantern, each of which consists of a glass catadioptric apparatus in front of the flame, with a silvered spherical reflector behind it. The radius of the glass apparatus is 75 millimètres, and is made to embrace 180°. It consists of a central refracting belt, with one prismatic ring above and below, and above and below the prismatic rings are two catadioptric prisms. The spherical reflectors are 8 inches in diameter, and are of silvered copper. The apparatus is hung in swinging gimbals, which allows of its remaining plumb during any movement of the ship under 30°.

The dioptric apparatus, adopted in the later lights, is the ordinary Fresnel design, with all recent improvements. The first order fixed apparatus has a cylindric refractor, consisting of a central belt with six prisms above and below it. The glass-work was put to the usual optical tests, by Messrs. Stevenson, before being taken off the manufacturer's hands.

The oil burned in the lighthouses has been, up to within a recent period, imported from China. The manufacture of oil in Japan is limited, and although some Japanese oils are good, no dependence can be placed upon getting a regular supply of any of them. One oil from China, procured from a nut, is a good burning oil, but its supply was uncertain, and its price often rose so high as to make its use almost prohibitive. Another Chinese oil, made from beans, is always to be had in large quantities, and its price is moderate, but its quality is variable and under the most favourable circumstances, it does not give satisfactory photometrical results. The supply of vegetable oils of the necessary quality being therefore doubtful, the Author deemed it advisable to recommend the adoption of mineral oils, and sanction was given to this in November 1872.

The burner invented by Captain Doty was selected. It is claimed for this burner, that it consumes a smaller quantity of mineral oil than the ordinary Argand burner does of vegetable oil, and that it produces a more powerful light, - that, 'volume for volume, mineral oil is superior to colza oil to the extent of one-fourth more light in the first order lamps, two-fifths more in the second, one-half more in the third, and four-fifths more in the fourth order lamps'; - that the flames are more easily attended to, and the standard heights of these more readily obtained; - that mineral oil is not affected by cold, and that it may be used without danger in hot weather; - that the same lamp-chimneys and wicks can be employed for burning mineral oil as for vegetable oil, and that the burners can be applied with ease to all existing lamps. Practical experience, afforded by the working of the Doty burner in Japan, bears out, to a great extent, these statements. The best result attained in an ordinary Argand lamp, with the oils formerly used, was a flame equal to ten standard sperm candles. There was always a difficulty in maintaining this standard, snuffing and trimming being necessary once, or sometimes twice, during the night. The Doty burner, in the same lamp, gives a flame equal to eighteen or twenty candles, which burns steadily, and generally without requiring any touching during the longest nights. The consumption of oil depends very much upon the skill of the lightkeeper. It is quite possible to cause an unnecessary consumption by the want of proper regulation of the ventilation in the lantern. The consumption has, therefore, varied considerably; but this difference will no doubt disappear as the lightkeepers get more accustomed to the use of the oil. The fitting of the Doty burner on the existing lamps was carried out with little expense or trouble. But the flame produced by a Doty burner, while exceedingly pure and brilliant, emits a much greater heat than that from a vegetable oil flame. The breakage of lamp-chimneys has on this account increased considerably. In the reflector lights of the first order, where there are in some cases as many as twenty-eight lamps in one lantern, the heat is, in certain states of the weather, very high and productive of great inconvenience. It is no uncommon thing for the temperature inside the lantern to be 130° Fahr., when that outside is 85°. This is the only real objection to the Doty burner in Japan, but there can be little doubt that it is capable of remedy by increased ventilation.

The mineral oil principally used up to the present time has been Young and Co.'s paraffin oil. It was procured direct from the works at Bathgate, and was imported in iron drums, each containing 5 gallons. Its cost, delivered in Yokohama, including all charges, was 75 cents per gallon. The specific gravity of this oil is 810, water being 1,000, and it does not emit any inflammable vapour until heated up to 145° or 150° Fahr.

Different samples of American kerosene have been tried. Its market price in Yokohama varies from 45 to 60 cents per gallon; but no specimens have been found of sufficiently good quality to warrant its use in the lighthouses. Its specific gravity ranges from 760 to 790, and many specimens had a flashing point as low as 80°. The best which was tried gave 120° as its flashing point, but it was consumed in a Doty burner, at 25 per cent. greater rapidity than the paraffin oil, and is altogether unreliable.

Petroleum exists in Japan in large quantities, and many wells have been sunk by the natives in different parts of the country. But no measures have yet been taken for carrying out the necessary refining processes, so as to enable it to be used either for domestic or for lighthouse purposes.

The principal office of the Lighthouse Department is at Yokohama. The workshops, store-rooms, &c., are erected on a space of about 4 acres of ground, and form a complete establishment, where the whole work required in the construction and maintenance of the lighthouses, lightships, and buoys, has been carried on. The yard is on the seashore, and is provided with a stone jetty, alongside of which boats can lie. On this is placed a crane capable of lifting 5 tons, and a line of rail runs from it to the different store-rooms.

In the yard there is an experimental lighthouse, a brick building, 20 feet square and 40 feet high. There are three floors, the lower one being used for testing oils, trying the power of flames, repairing or readjusting apparatus, or doing such other experimental work as may be required. The second and third floors are used for the purposes of the lighthouse. The lantern was entirely constructed in the lighthouse yard, and is of the ordinary size for first order lanterns. The different floors are fitted with trapdoors, sufficiently large for the largest apparatus to pass without being taken to pieces. The principal object of the lighthouse is to afford means for furnishing a preparatory training to young Japanese lightkeepers previous to their being sent to the regular lighthouse stations. It is also used for examining apparatus, which during its voyage from England may have got shaken out of adjustment, or have been otherwise damaged. The mineral oil store, built of concrete, with walls 2 feet thick, and an arched roof of brick, is capable of containing 30,000 gallons of oil, in the boxes in which it usually comes from England. There is also a vegetable oil refinery, with the boilers, agitators, and vessels necessary for settling and filtering the oil. Everything connected with the lighthouses, with the exception of the optical apparatus and lanterns, has been constructed in the establishment. The two lightships were built, fitted out, and entirely completed in the yard; twenty-two buoys were made, and the iron and wooden lighthouses were fitted previous to being sent to their destinations.

The whole work has been carried on with the help of two assistant engineers, Mr. James McRitchie, Assoc. Inst. C.E., and Mr. S. Fisher. One secretary and two clerks were also attached to the establishment. The other Europeans engaged in the service at different times were: one superintendent blacksmith, two inspecting and repairing artisans for the apparatus, one plumber and coppersmith, one shipwright, four masons, and fifteen lightkeepers.

The Japanese staff is presided over by a commissioner, with two assistant commissioners acting under him. These are granted certain powers by the Public Works Department of the Government, and they are supposed to work in concert with the chief Europeans in the office. But, in point of fact, they are vested with the supreme control, and the Europeans have little or no authority. Under the commissioners there is a staff of fifty-three Japanese officials, who are divided among four departments, viz.: the secretary's, the accountant's, the building or construction, and the storekeeper's departments. The organisation of the office was arranged so as to serve the peculiar requirements of the service. The Author compiled, from English, French, and American sources, one book of instructions for lightkeepers and another for repairing artisans. These were translated into Japanese; and are closely adhered to. The former gives minute directions regarding every process in connection with the lights, and orders meteorological records to be kept at each station. The latter explains the adjustments necessary in the apparatus, and the manner in which repairs are to be carried out.

The European lightkeepers are distributed among the most important ocean lighthouses, but their numbers are being gradually reduced, as the Japanese become more expert and efficient. The Japanese lightkeepers number one hundred and two, but only a few of them are sufficiently advanced to have the entire charge of a light, the remainder acting as assistants under the Europeans. There are generally four Japanese at each station under the European, one of whom is principal, the remaining three being pupils in the course of training. The European lightkeepers have authority to demand obedience from the Japanese in any matter connected with the keeping of the lights, or the general good order of the station, but they have no further control over them; and the execution of repairs, procuring stores, or the making of any payments is done through the Japanese principal lightkeeper. The Europeans are specially directed to be attentive in instructing the Japanese lightkeepers, and rewards are held out to those who succeed in training the most efficient men.

Each lighthouse is inspected, on an average, three times every year by one of the Japanese commissioners, accompanied by one of the engineers. The only ready means of communication with most of the lighthouses is by sea; and the vessel procured for carrying on the work of inspection is a paddle-wheel steamship of 489 tons register and 370 nominal HP. She is easily handled in confined harbours, of great power, and capable of steaming out of danger under any circumstances, with a capacious deck for buoys, and the necessary passenger and cargo accommodation. She is commanded and officered by Europeans, but her crew consists of Japanese only.

Owing to the manner in which the Japanese keep accounts, and their desire to have the whole control of the expenditure, the Author, notwithstanding the utmost efforts, has failed to preserve a correct record of the cost of the works. For the first year a system of certificates was instituted, by means of which he had a check upon all moneys paid; but he soon discovered that many payments were made by the Japanese officials without any certificate, and the practice had to be abandoned as useless. He has further been in partial ignorance of the prices paid for material and labour. The statement, in Appendix II, of the money expended in the construction and maintenance, of each lighthouse, has chiefly been supplied by the Japanese officials.

The Author arrived in Japan just at the completion of the civil war, which revolutionised the country and gave Europeans a more secure footing in it. He then found the various Government departments in a disorganised state. The lighthouse work was transferred from one to another, and at various times was in charge of about eight or ten different sets of officials. It so became difficult to organise the service, or to arrange or enforce any rules or regulations. At the time of his arrival no artisans were acquainted with anything beyond the slight and unimportant work customary in the country. Carpenters were skilful in the use of their tools, but masons, bricklayers, and blacksmiths were almost unknown. Men had therefore to be taught each of those trades, and the works had to be carefully watched during their progress. The staff of Europeans employed has been extremely small, but notwithstanding the difficulties surrounding the prosecution of the works, each lighthouse is well finished. There have been no signs of failure of any importance, and the earthquakes which have occurred since their erection - one or two were severe - have had no visible effect upon them.

The Author has received most flattering assurances of the high appreciation in which the lights are held by the nautical men who frequent the coasts of Japan; and great credit is due to the Japanese Government for the liberal manner in which it has prosecuted the work.

The Paper is accompanied by a series of drawings, from which Plates I and 2 have been compiled.

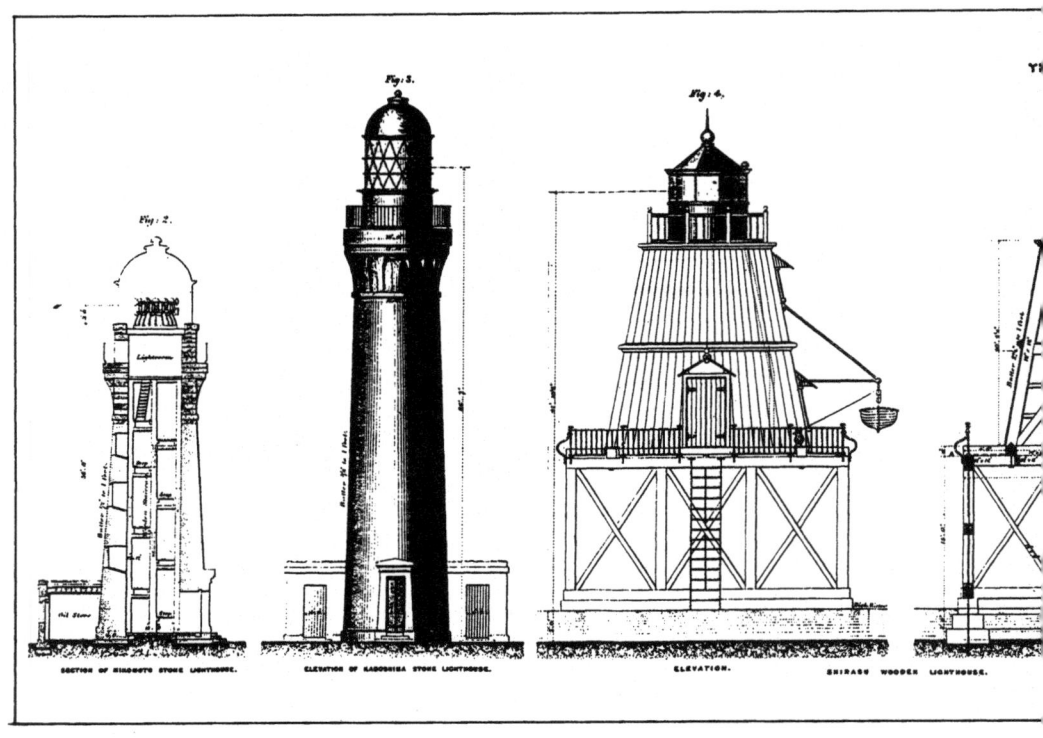

Fig: 2. *Fig: 3.* *Fig: 4.*

SECTION OF HIROOTO STONE LIGHTHOUSE.   ELEVATION OF KAGOSHIMA STONE LIGHTHOUSE.   ELEVATION.   SHIRASU WOODEN LIGHTHOUSE.

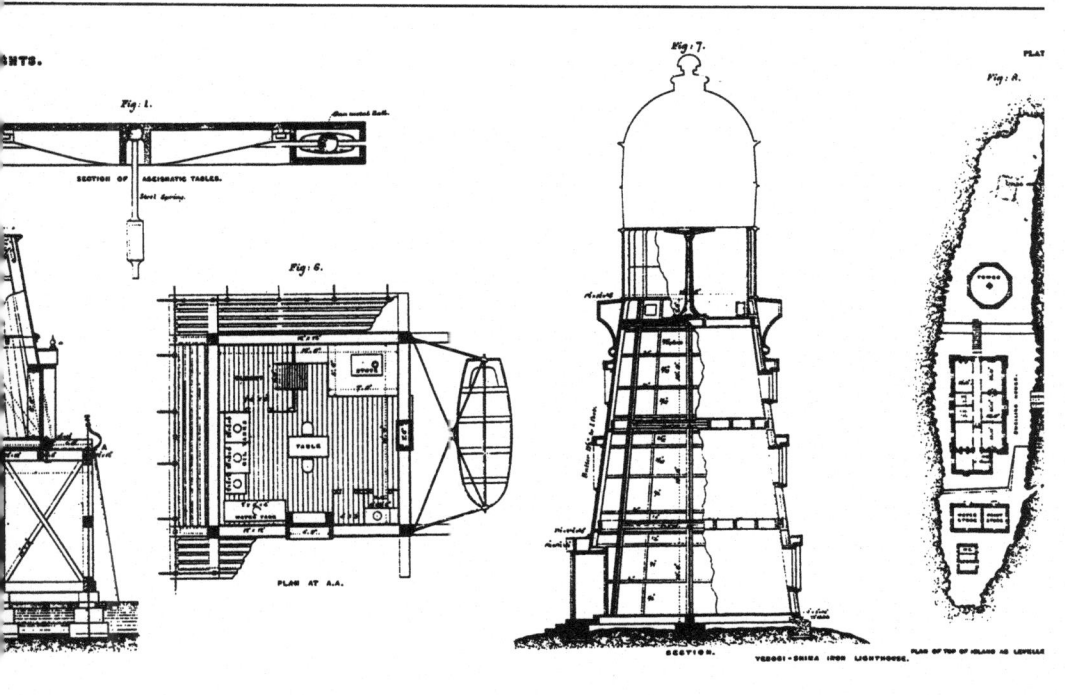

Fig: 1.

SECTION OF AERIGRATIC TABLES.

Fig: 6.

PLAN AT A.A.

Fig: 7.

SECTION.

YEBOSI-SHIMA IRON LIGHTHOUSE.

PLAT

Fig: A.

PLAN OF TOP OF ISLAND AS LEVELLED

LIST OF LIGHTHOUSES, LIGHTSHIPS, BUOYS AND BEACONS ON THE COAST OF JAPAN

LIGHTHOUSES

1	Shinagawa	Dioptric	Red	Fixed	5th Order
2	Yokohama Harbour	,,	,,	,,	Staff light
3	Haneda	,,	Green	,,	4th Order
4	Kannonsaki	,,	{ White and Red	,,	3rd ,,
5	Tsurugisaki	Catoptric	,,	Flashing	2nd ,,
6	Noshimasaki	Dioptric	White	Fixed	1st ,,
7	Inuboyesaki	,,	,,	Revolving	1st ,,
8	Kinkasan	,,	,,	Fixed	1st ,,
9	Siriyasaki	,,	,,	,,	2nd ,,
10	Awomori	,,	Red	,,	Staff light
11	Ishonomaki	,,	White	,,	,,
12	Noshiaf	,,	,,	,,	,,
13	Nemuro	,,	Red	,,	,,
14	Jokashima	,,	White	,,	5th Order
15	Mikomoto (Plate 2, Fig. 2)	Catoptric	{ White and Red	,,	1st ,,
16	Irosaki	,,	Red	,,	6th ,,
17	Omaesaki	Dioptric	White	Revolving	1st ,,
18	Sugashima	,,	,,	Fixed	4th ,,
19	Matoya	,,	,,	Revolving	4th ,,
20	Kashinosaki	Catoptric	,,	,,	2nd ,,
21	Shiwomisaki	,,	,,	Fixed	1st ,,
22	Tomagaishima	Dioptric	,,	,,	3rd ,,
23	Temposan	,,	,,	,,	4th ,,
24	Wadanomisaki	,,	Red	,,	4th ,,
25	Yesaki	Catoptric	White	,,	1st ,,
26	Nabeshima	Dioptric	,,	,,	3rd ,,
27	Tsurishima	,,	,,	,,	,,
28	Hesaki	,,	{ White and Red	,,	4th Order
29	Rockuren	,,	White	,,	,,
30	Kadoshima (Plate 2, Fig. 3)	Dioptric	White	Flashing	1st Order
31	Shirasu (Plate 2, Figs. 4, 5 and 6)	,,	Red	Fixed	5th ,,
32	Yebosishima (Plate 2, Figs. 7 and 8)	,,	White	,,	2nd ,,
33	Iwoshima	Catoptric	,,	,,	1st ,,
34	Satanomisaki	,,	,,	,,	,,

35	Yokohama Bay	Catoptric	Red	Fixed
36	Hakodate Harbour	,,	White	,,

Of these thirty-six lights there are ten first-order, four second-order, four third-order, seven fourth-order, three fifth-order, one sixth-order, five staff lights, and two lightships.

BUOYS

	Height in Feet	Positions
1 Can buoy	8	South of Yokohama Harbour
2 ,, ,,	8	North of ,,
3 ,, ,,	8	On Kawasaki Spit, Gulf of Yedo
4 ,, ,,	8	On ,, ,,
5 Nun ,,	15	On Saratoga Spit, Gulf of Yedo
6 Can ,,	8	On Kanabuse, Shimonoseki Straits
7 ,, ,,	8	On Middle Ground, ,,
8 ,, ,,	8	On ,, ,,
9 Hollow-bottomed buoy	9	At Hikushima, ,,
10 ,, ,,	9	,, ,,
11 ,, ,,	12	At Hiraiso, near Awaji Sima
12 ,, ,,	12	At Motoyama in Inland Sea
13 ,, ,,	12	North of Island of Ainoshima, west of Shimonoseki Straits

BEACONS

1 Stone beacon	20	On Manaita Rock, Shimonoseki Straits
2 ,, ,,	20	On Narusi Rock, ,,
3 ,, ,,	20	On Yodsibi Rock, ,,

APPENDIX II

EXPENDITURE INCURRED BY THE LIGHTHOUSE DEPARTMENT FROM THE DATE OF ITS ESTABLISHMENT, AUGUST–1868, TO THE END OF 1875.

Name of Light	Date of Commencement	Date of Completion	Height of Tower of centre of Lantern	Material	Cost
			Feet		Dollars
1 Shinagawa	Feb. 1873	Apr. 1873	19	Brick	2,054
2 Yokohama Harbour	Jan. 1874	Mar. 1874	40	Wood	245

3	Haneda	Mar. 1874	Mar. 1875	55	Iron	14,000
4	Kannonsaki	Oct. 1868	Dec. 1868	40	Brick	8,182
5	Tsurugisaki	Mar. 1870	Mar. 1871	25	Stone	34,444
6	Noshimasaki	Mar. 1869	Dec. 1870	99	Brick	22,599
7	Inuboyesaki	Feb. 1872	Nov. 1874	90	,,	44,825
8	Kinkasan	Mar. 1874		24	Stone	45,100
9	Siriyasaki	June 1873		92	Brick	61,300
10	Awomori	July 1874	Nov. 1874	40	Wood	424
11	Ishinomaki	Dec. 1873	Feb. 1874	40	,,	424
12	Noshiaf	June 1872	Aug. 1872	31	,,	2,519
13	Nemuro	May 1872	Aug. 1872	40	,,	424
14	Jokashima	July 1870	Sept. 1870	19½	Brick	2,090
15	Mikomoto	Apr. 1869	Jan. 1871	60	Stone	107,530
16	Irosaki	July 1871	Oct. 1871	20	Wood	2,430
17	Omaesaki	June 1872	May 1874	57	Brick	33,340
18	Sugashima	Feb. 1872	July 1873	28½	,,	8,912
19	Matoya	Oct. 1871	Sept. 1872	36	Wood	13,023
20	Kashinosaki	Apr. 1869	Aug. 1870	15	Stone	40,229
21	Shiwomisaki	Apr. 1869	Oct. 1870	63	Wood	50,253
22	Tomagaishima	May 1870	July 1872	21	Stone	22,963
23	Temposan	Dec. 1870	May 1871	30	Wood	6,509
24	Wadanomisaki	Oct. 1870	May 1871	46	,,	13,517
25	Yesaki	May 1870	May 1871	15	Stone	28,119
26	Nabeshima	Dec. 1871	Feb. 1873	21	,,	7,919

Name of Light	Date of Commencement	Date of Completion	Height of Tower to centre of Lantern	Material	Cost
			Feet		Dollars
27 Tsurishima	Oct. 1871	July 1873	21	Stone	18,823
28 Hesaki	Dec. 1870	Apr. 1872	21	,,	20,204
29 Rockuren	Dec. 1870	Dec. 1871	25	,,	23,710
30 Kadoshima	Aug. 1873		87	,,	52,000
31 Shirasu	Mar. 1872	Mar. 1872	42	Wood	9,239
32 Yebosishima	Aug. 1873	Aug. 1875	44	Iron	69,400
33 Iwoshima	June 1869	Jan. 1872	23	,,	21,920
34 Satanomisaki	Jan. 1870	Dec. 1871	20	,,	66,838

LIGHTSHIPS

35 Yokohama Bay	Dec. 1868	Dec. 1869		Wood	29,645
36 Hakodate Harbour	May 1870	Apr. 1871		,,	29,260

BUOYS (including spare ones)

	Buoys	Height in feet	
Yokohama	3	8	2,379
Kawasaki	3	8	2,379
Saratoga Spit	2	15	3,320
Shimonoseki	5	8	3,965
Hikushima	3	9	2,370
Hiraiso	2	12	2,433

Motoyama	2	12	2,433
Ainoshima	2	12	2,433
Total	22		

BEACONS

Manaita ⎫
Narusi ⎬ Commenced April 1871. Completed September 1871 1,527
Yodsibi ⎭

Lighthouse establishement, offices, and dwellings	66,237
Total cost of construction	$1,003,889

The purchase of the steam tender and her maintenance, the payments of all salaries to both European and Japanese officials, and various other miscellaneous expenses connected with the establishment amount to   1,050,000

The maintenance of the various lighthouses amounts to   189,000

Making a grand total of $2,242,889

Or about £450,000 laid out on the lighthouse-work up to the end of the year 1875.

Messrs. STEVENSON, having been intrusted by the Board of Trade to advise as to establishing a lighthouse system for the coasts of Japan, and to select and recommend for appointment engineers, foremen-mechanics, and trained lightkeepers to carry out their recommendations, desired, through the Secretary, to acknowledge the valuable services of Mr. Brunton, who had been selected for the office of Chief Engineer, in carrying out their designs. The Author of the Paper had not adverted to certain preliminary matters (probably because they occurred in 1867, before his appointment), which explained correctly what led to the adoption in the first instance of catadioptric instead of dioptric apparatus, and rendered it unnecessary to offer any observations on what the Author had said as to the comparative advantages of the dioptric and catadioptric systems of illumination, as to which lighthouse engineers did not differ in opinion.

These preliminary matters referred to the authenticated effects of earthquake shocks, to which attention had been drawn in all the reports sent from Japan to the British Government. In transmitting these reports to the Messrs. Stevenson, the Board of Trade specially requested them to consider in what way the anticipated injury to the apparatus could be ameliorated. The Elder Brethren of the Trinity House, who had also been consulted by the Board of Trade, suggested the employment of Holophotal reflectors as the safest apparatus that could, under the circumstances, be employed, a recommendation in which Messrs. Stevenson entirely concurred. As an additional safeguard they proposed the introduction of an 'aseismatic' arrangement. This, as applied to the apparatus, consisted of a slight alteration of the table on which the reflector frame and machine rest, at a cost of about £90 for each lighthouse.[3] Messrs. Stevenson were confirmed in their recommendation of this simple expedient from satisfactory experiments made in this country, which proved that violent shocks applied to the lower part of the table, failed to affect apparatus resting on it, so long as the aseismatic joint was in operation; but that whenever the table was screwed down, so as to form a rigid mass, the same shocks not only threw off the lamp glasses, but in some cases extinguished the lights. Fig. 1 was a section in which the upper table bearing the holophotal reflectors was shown in hatched lines. This table rested on balls contained in cups formed in the upper and lower tables. Fig. 2 was a plan showing the points of support which, for ease of adjustment, had been made at three places dividing the circumference of the table into three equal arcs. The same arrangement had been proposed on a larger scale for the foundations of conical towers of plate iron. The vessel containing these towers was unfortunately wrecked, and no experience of their action existed.

In further illustration of the subject, and particularly of what had been done in Japan, Messrs. Stevenson gave the following quotation from their Report to the Japanese Government made in June 1876, after the authorised works had been completed:[4] 'In all of the documents,' the report states, 'which were submitted to us for consideration by the Board of Trade in 1867, whether from the French, American, or English authorities, great prominence was given to the prevalence of earthquakes, and their anticipated effects on the proposed lighthouse works. Captain Bullock, R.N., calls attention to the frequent volcanic shocks, and says, "Engineering skill will be required to provide against them." The French naval officers report that earthquakes, so frequent in Japan, "would prevent the use of stone structures," and the Trinity House report "that perhaps the greatest difficulty of all will be found in the constant volcanic disturbances common in Japan." To meet the difficulty which had been raised Mr. David Stevenson suggested the use of an "aseismatic joint" in the table supporting the lighting apparatus, so contrived as to admit of a certain amount of horizontal movement taking place in the lighthouse without affecting the table on which the apparatus rests. This arrangement was proved on experimental trials made in this country to be successful, and it was introduced into the Japanese lighthouse apparatus. It would appear, however, that from some dislike to the unsteadiness of the apparatus during the process of cleaning, the aseismatic joint was secured so as to prevent it from acting; and, with the apparatus in this state, the only recorded earthquake which has affected any of the lights occurred at Segami, which is lighted by twenty-one holophotal reflectors. The shock had the effect of throwing the whole of the lamp-glasses out of their places, and disarranging the working of the machine, and the opportunity of testing the effect of the apparatus (as an amelioration of an earthquake shock) was thus lost. The experience obtained since the lighthouses were established has certainly gone to prove, that earthquake shocks have neither been so frequent nor severe as was represented; but it should be kept in view that such visitations are extremely uncertain, and may return at any time with renewed frequency and increased vigour. We should, therefore, suggest that if it be desirable to make the tables *less sensitive*, that this could easily be arranged without rendering them incapable of motion; or if it be considered desirable to screw up the tables while the apparatus is being cleaned, the screws should be made so as to be undone when that operation is completed, so that the responsibility for the action of the aseismatic apparatus should lie with the lightkeeper and not with the lighthouse authorities. The serious damage of any of the lighting apparatus by earthquake shocks, would entail undesirable responsibility if it should occur in a case where the aseismatic joint was found to have been permanently screwed up so as to prevent its acting.'

Messrs. Stevenson had great pleasure in saying how much they were assisted by the valuable advice of Admiral Bedford, of the Board of Trade, in determining the positions and the characteristics of the most important of the sea lights; as well as in arranging the various details for giving the Japanese authorities the aid asked from the Government of this country in establishing their lighthouse system.

Mr. DOUGLASS observed that, this being a case of the first lighting of a large extent of coast line, he should be glad to hear some explanation as to the large proportion of fixed to revolving or flashing lights. It was well known that revolving or flashing lights had an accumulative power from five to ten times greater than fixed lights, with the same consumption of oil; in other words, the annual cost for oil with a revolving or flashing light was from one-fifth to one-tenth of that of a fixed light of the same power. In first lighting a coast like Japan, an essential consideration would, therefore, seem to be to introduce as many revolving or flashing lights as possible, having due regard for the necessary distinctions to prevent one light being mistaken for another. The fixed light was considered to be the least distinctive, and should only be adopted where it was desirable to use coloured arcs to cover local dangers; but that appeared, from the Paper, to have been done only in three cases. Out of thirty-four lights only five were revolving. He thought that, as a provision against earthquakes, a light wrought iron, or wooden, structure would be preferable to one of stone. If the stone structure were low, as in some of the instances described, it might, no doubt, withstand a heavy shock; but, for a tower 80 or 100 feet high, a light iron structure, or one constructed of local timber, would have been less costly and more stable. If the earthquake shocks in Japan were as serious as had been represented, he should prefer the old catoptric revolving light. With this light the consumption of oil for the same amount of light would be greater than with a dioptric revolving light, but less than with a dioptric fixed light, while the first cost of the apparatus would be much less than that of a dioptric light, either revolving or fixed. With earthquake shocks these lights would be as safe as any building that an engineer could design. Evidence of this had been furnished by the forty-seven light-vessels in the service of the Trinity House. These vessels were all fitted on the catoptric system. In some collisions, the lantern and illuminating apparatus had fallen from the mast-head to the deck, a height of 30 feet, but in no instance had the apparatus been entirely destroyed. Under such circumstances, a dioptric apparatus would have been completely broken to pieces. A fixed catoptric apparatus was almost out of the question except where very small arcs were required to be illuminated, because it was impossible to produce a well-defined cut between white and coloured arcs. It was stated in the Paper, that in the Doty burner, 'volume for volume, mineral oil is superior to colza oil to the extent of one-fourth more light in the first order lamps, two-fifths more in the second, one-half more in the third, and four-fifths more in the fourth order lamps.'[5] From his experience with the Trinity House lamps - with which the highest recorded photometric results with the greatest economy of working had been obtained - he had found that volume for volume there was no superiority in the best mineral oil over the best colza, but rather a small percentage in favour of colza; moreover, with colza oil these lamps burned almost uniformly for sixteen hours, with only the usual occasional regulation of the damper, but without any trimming of the wicks. The only known advantages of mineral oil over the best vegetable oils, including cocoa-nut, which was as good as colz, were its readiness of ignition, its flame more rapidly rising to the maximum power, and in some cases its low price - the present price in this country being about two-thirds that of colza.

Mr. W. LLOYD said, having had some experience in the construction of lofty buildings in countries liable to earthquakes, he thought there was a great deal more fear with regard to the effects of those shocks than was at all justified. He should like to give two instances that had come under his own observation bearing on the movable joint that had been proposed. As he understood it, this would allow of a limited horizontal movement, and of a free vertical movement. The first case was that of a church-tower in Mendoza, which had been almost entirely destroyed in the year 1861. Two storeys out of three remained after the earthquake. The lower storey was more or less intact, except that there was a large crack in it. The upper storey was entirely destroyed; but the middle storey, by some horizontal movement, was turned round nearly one-half upon the lower storey, so that the angles of the upper storey corresponded more or less with the centre of the side of the lower storey. That showed an amount of horizontal movement which would be dangerous with a joint of the description proposed. With regard to the vertical movement in the case of an earthquake, he would cite an instance that occurred in Valparaiso in 1851. A family party were sitting in a room after dinner, and on the table was one of the old-fashioned oil lamps, with a large globe and a chimney-glass within it. An earthquake took place, the family rushed out, and on their return they found the globe on the ground; it had been thrown over the chimney-glass, which remained in its place. One could easily imagine, therefore, what the effect would be in a lighthouse, if the joint had a vertical movement. The lantern might be destroyed. From his experience of the structures erected by the Spaniards, he believed that the more solid the structures, the better they resisted earthquake shocks. In Panama there was a single-stone arch of an old Spanish church, spanning 40 or 50 feet, and it remained to the present day. On the Pacific coast, numerous bridges had been built by the Spaniards with solid stone arches, which had resisted every earthquake, not a crack being visible in them. He believed that Nature had pointed out the best course to pursue in structures of that kind. In the lighthouses described, so far as he could judge from the illustrations, there was one defect - a want of batter. Every structure that he had erected in an earthquake country had a great deal more batter than would be given to a similar structure in England. The broader the base the better. An analogy might be found in the buttress-tree, which spread out on all sides in a manner remarkably like buttresses.

Dr. POLE could testify, from correspondence he had had with Japan, how highly Mr. Brunton himself and his work were valued by the Japanese and by his brother engineers in that country. He had placed upon the wall a Japanese map on a large scale, which had been recently sent to him by Mr. Boyle, the Chief Resident Engineer of the railways, who had marked upon it the lines executed and projected. He thought, at some future time, a Paper upon Japanese railways might be interesting to the members of the Institution. He had also placed upon the table some photographs of railway works in Japan. The country was no doubt a land of earthquakes. Near Yokohama there was a celebrated conical mountain, Fusi Yama, always seen in Japanese lacquer-work, very high, and visible at a long distance. It was a volcano, which had been in action within the historical era, but it was now extinct. The volcanic influence, however, remained, and the country, especially in that particular neighbourhood, was very subject to earthquakes. He need not say that that liability had attracted the attention of the railway authorities, who had endeavoured to make their works strong enough to withstand the shocks. A great many bridges were required, and many of them had been constructed of iron sent out from England. In their design the probable effects of earthquakes had also been considered. But there had been no shocks of importance since the railways had been opened, and experience had therefore yet to be gained.

Mr. ABERNETHY, Vice-President, observed that probably there was something special in the character of the Inland Sea, otherwise he could not agree to the principle laid down by the Author, to put no lights where the headlands were bold and well defined, or where no danger existed. In this country the most important class of lights were those placed upon salient headlands, in order that during on-shore gales vessels might have a good offing, and ascertain their proximity to land as early as possible. The system of lighthouses might be divided in three parts: 1, the lighthouse upon prominent headlands; 2, lights to define the position of harbours; and 3, guiding and leading lights to guide vessels into harbour. He could not gather from the Paper that this order had been observed. With reference to the effect of earthquakes, he was of opinion that weight and solidity, and not slightness of structure, were best adapted to meet shocks of that description. An extraordinary example was that of the great aqueduct from Cintra to Lisbon, which at one point was at least 250 feet in height, and which was within the zone of the great earthquake that destroyed nearly all the principal buildings, when this structure was uninjured and remained intact to the present day. This fact among others led him to believe that, in order to meet the shock of earthquakes, solidity and weight were essential elements.

Mr. WOODS said he had been engaged in the construction of buildings for Peru, where earthquakes were very prevalent, and he had always understood, from the information he had received, that buildings of slight construction, whether of wood or of iron framing, were better adapted to resist earthquake shocks than more solid erections. He could mention cases in which light structures had stood very severe earthquake shocks. For instance, at Pisco, a small town on the coast of Peru, exposed to the action of the sea with heavy surf, there had been built from his designs, about twenty-five years ago, a light wrought-iron pier, with screw piles, extending 2,400 feet out to sea. Six or seven years ago a great earthquake occurred, by which the city of Arica was overwhelmed. A wave of the sea rushed into the town, carrying with it a gunbrig ½ mile inshore, where she now lay high and dry. The pier sustained the shock of the earthquake. The motion of the platform was so severe, that persons could not stand steadily upon it, but the pier suffered no injury whatever, the undulations passing through it without affecting its stability in any way. He believed the pier was now as sound as when it was first erected. He might mention another instance, that of a building erected at Payta, also on the coast of Peru. It had a light iron framework, lined with wood, and it had been in existence twenty-five or thirty years, after sustaining every shock that had occurred in that part of the country. He did not know whether there had been sufficient experience to test the effects of earthquakes on the solid lighthouses in Japan, but he was inclined to think that a severe shock might occasion them considerable disturbance. With regard to the proposal to fit the lighthouse with lantern tables, he thought an elastic fixing would be as serviceable as the method described in the Paper. If the platform to receive the lanterns were secured to a proper table, by means of intervening springs, he believed all the effect supposed to be obtained by a ball and socket joint would be obtained.

Mr. BEAUMONT said with reference to the partial revolution of one portion of a building upon another, it had been shown by Mallet, in his various works on the phenomena of earthquakes, that it was due to the difference between the frictional adhesion of one part of the building and that of the other, the inertia of the building allowing that part that was not as well fastened as the other to remain nearly still while the other part or parts moved backwards or forwards in the first or second phase of the shock.[6] In the case of lighthouse structures, where the different courses were properly dovetailed one into another, of course that would not take place; neither would structures built of iron be subject to any such movements.

Mr. BRUNTON, in reply, said one reason for the large proportion of fixed lights was their greater simplicity, a matter of importance in a country like Japan; but it should be understood that only part of the scheme for the illumination of the coast had at present been carried out; in order to introduce the necessary distinctions, the lights still to be established would be chiefly revolving. The natives would by that time have gained experience, and might be intrusted with a revolving machine; but, at the commencement of the undertaking, the importance of having a simple apparatus could not be exaggerated.

Doty's burner was adopted in Japan in 1872, on account of the difficulty experienced in procuring suitable vegetable oils, and because at the time no other suitable mineral oil burner was available. Subsequent improvements in burners by Mr. Douglass tended to show that good colza oil could be consumed to give equally as good results, photometrically, as mineral oil; but this was not universally admitted. The standard power of each size of flame, fixed for the purpose of testing the oils used in Japan, and the standard consumption to attain these powers, were as follows:-

	Power	Rate of Consumption of 1 Gallon
4-wick colza oil lamp	= 255 candles	4¾ hours
4-wick mineral ,,	= 270 ,,	5 ,,
3-wick colza ,,	= 150 ,,	9 ,,
3-wick mineral ,,	= 190 ,,	9¼ ,,
2-wick colza ,,	= 50 ,,	17 ,,
2-wick mineral ,,	= 75 ,,	17½ ,,
1-wick colza ,,	= 10 ,,	86½ ,,
1-wick mineral ,,	= 19 ,,	92 ,,

Flames of these strengths were invariably maintained in the Japan lighthouses. These figures were the result of repeated, independent experiments. Allowing that only 23 per cent. more heat was evolved by mineral oil, yet the accumulation of that extra heat in a lantern during a calm tropical night was, in his opinion, sufficient to produce the effects he had stated.

The Japan Lights had been thus classified in the Paper:- 1, Lights required by the treaty powers, which were principally ocean lights; 2, Lights for the Inland Sea; 3, Local lights, these being mostly harbour lights; 4, Ocean lights established by the Government. These, in his opinion, included the three divisions named by Mr. Abernethy. He had not advocated as a general principle 'to put no light where the headlands were bold and well defined, or where no danger existed';[7] but it was one which had been adopted, after many consultations with nautical men, in reference solely to the lighting of the Inland Sea, where the circumstances were peculiar.

The Paper was necessarily of a general character, because it described a system of lighthouses and not any one in particular; but as many details of construction had been given as were deemed to be of interest. He had consulted all available authorities regarding the nature of the structures best calculated to resist earthquakes; and there seemed to be a general approval of solidity, weight, and strength, as opposed to lightness and flexibility. He accordingly adopted that principle. A lighthouse was a simple erection, and he had given in the Paper the form of each, the thickness of the walls, the sizes of the stones, and the manner in which they were arranged. The batter of the walls was in every case considerably greater than would have been thought necessary in this country.

Many violent earthquakes had occurred since the lighthouses had been erected. But, on account of the inconveniences occasioned by the unsteadiness of the upper part of the aseismatic arrangement, as designed by Messrs. Stevenson, it was found necessary to secure it, and so prevent its acting. From a desire to give the design as full a trial as possible, he had repeatedly given instructions to lightkeepers at different stations to keep the tables free to act, and to report to him the result. In every instance their reports showed that the movements caused by the operation of trimming the lamps, or of performing the various other duties connected with the light - and in an apparatus having twenty-eight separate burners these were frequent - seriously deranged the flames, and prevented the machinery in revolving lights working smoothly. Under these circumstances, it became, in his opinion, imperative to prevent the tables acting. The opinion of the Messrs. Stevenson, that the tables might easily be made less sensitive without rendering them incapable of motion, was precisely the problem still to be worked out. The degree of sensitiveness to be attained was one which, while unaffected by ordinary disturbances, would afford a free motion in earthquakes. Probably to attain this might present some difficulty. An earthquake had recently displaced the glass cylinders of the twenty-one reflector lamps at Tsurugisaki lighthouse. These merely stood on their bases, and were not supported in any way. This was the only damage done at the time in the lighthouse. The earthquake, though apparently not so severe as some previously experienced, had a peculiar motion. It turned the upper six or seven courses of the stonework of one of the chimneys of the keeper's house - a column of ashlar masonry about 2 feet by 3 feet, and 10 feet high - half round, and left it in a diagonal position to the lower part, without other fracture or displacement. Mr. Mallet, in his 'History of the Neapolitan Earthquake of 1857', and Mr. Lloyd, both mentioned that they had observed similar twisting movements. In close proximity to Tsurugisaki there were three dioptrical lights, viz., Noshima, Kannonsaki, and Jokashima, where the earthquake had been as severely felt. In these, the large glass cylinders for the single flames were secured both at the top and at the bottom, but no derangement whatever had occurred. So far as the experience - which was not very extensive, however, in Japan - went, dioptric apparatus was as little liable to derangement from earthquakes as catoptric. Mr. David Stevenson, in his Paper read before the Royal Scottish Society of Arts, said that any sudden lateral motion of the earth on which a building rests, must be transmitted through the rigid and unyielding material of the structure to the summit, where the violence of the shock would be aggravated. Having regard to the vibratory character of the movement in an earthquake, the motion being as much backward as forward; and also to the fact that no building material was absolutely rigid, or unyielding, it was clear that a shock would gradually lose in intensity as it left the foundation and approached the summit of a structure. As a matter of fact, earthquakes, which had been sufficiently severe to frighten the lightkeepers and their families out of their dwellings, had not been felt at the summits of the higher lighthouse towers in Japan. This was of great importance in the construction of buildings in earthquake countries. Recent destructive earthquakes in Japan had invariably indicated a vertical motion only; and the destruction to erections had occurred on the sharp downward movement which always followed the upheaval. Messrs. Stevenson's aseismatic tables did not provide for such a movement.

The Author begged to acknowledge the facilities which were afforded him by the Messrs. Stevenson, before he proceeded to Japan, for obtaining a detailed and accurate knowledge of lighthouse construction. The records of their office were placed at his disposal, and they personally assisted him in every way, especially as regarded the design and construction of the optical apparatus, in which matters they were pre-eminent. He also mentiond the assistance he had received at one time from H.E. Mr. Wooyeno, Minister in England for Japan, who, almost at the outset, was the commissioner appointed by the Government to take charge of the work. By his energy and high intelligence he succeeded in extricating the work from serious difficulties in which it was involved, owing to the peculiarities of the native workmen. Unfortunately his Excellency only remained in the Lighthouse Department for a short time, by which the assistance of a willing and an able colleague was lost, just at the time when such services were most needed.

Mr. W. D. CAY referred, through the Secretary, to the advantages of the use of gas for lighthouses in situations occasionally inaccessible. He had recently completed the works for lighting, in this manner, the lighthouse at the seaward extremity of the New South Breakwater at Aberdeen, the tower of which had been described in the Minutes of Proceedings of the Institution.[8] The lightkeeper was unable to pass along that breakwater during storms on account of the waves, so that it was necessary to have a constant light; and it had been so arranged that this was kept low during the day, and the full pressure was turned on at sunset by a cock placed on the shore. The gas was brought from the town of Aberdeen in a 2-inch pipe, upwards of 2½ miles long, crossing the river Dee by the Suspension Bridge in a lead pipe, and was carried along the breakwater in a wrought iron pipe sunk in a groove cut in the concrete, and covered with a pitch-pine beam. A gasholder, inclosed in a house, was provided at the shore end of the breakwater, holding a week's supply of gas; and a dry gasmeter, placed on the pipe from the town, prevented the return of gas towards the town, when through any cause the pressure in the supply pipe should be less than the pressure in the holder. These arrangements had been completed, and tested, and had been found to work in a satisfactory manner. The total cost, including the lighthouse which had been built of concrete in 1873, the lightroom, lanterns, gas supply works, &c., would be about £2,200.

Mr. J. J. COLEMAN remarked, through the Secretary, that it was only within the last few years that mineral oil had been used for lighthouse purposes. Previous to 1868 the possibility of using mineral oil in lighthouses had been considered by the Trinity House, but the then existing lamps were not found suitable for its application. In the month of March of that year Captain Doty submitted to the Trinity House authorities a burner with a concentric wick, which he asserted would answer the purpose. This, however, was a crude affair; and the Deputy Master lent Captain Doty a lamp of the first order as a model. Another and more perfect burner was constructed, and sent in the course of the same year to the Trinity House. Professor Tyndall examined it, and did not advise its adoption; but, during the two ensuing years experiments were made by Mr. J. N. Douglass, M.Inst.C.E., who altered the existing Trinity House burners, so as to adapt them for burning mineral oils. A first order light so altered was used successfully at Flamborough Head in 1871.[9] In the meantime, Captain Doty had succeeded in introducing his burner, and the use of mineral oil for lighthouse purposes, into France; the late Emperor of the French taking considerable interest in the subject. He was also successful in interesting the Commissioners of Northern Lighthouses, through their

Engineers, Messrs. D. and T. Stevenson. Dr. Stevenson Macadam examined the photogenic power of mineral oil when burnt in Captain Doty's lamp, and made an exhaustive report to these gentleman on the 31st of December 1870. Assuming a given weight of colza oil to yield a light denoted by 1,000, Dr Macadam stated that an equal weight of mineral oil furnished in

First order of lamp, a light equivalent to	1,184
Second ,, ,, ,,	1,544
Third ,, ,, ,,	1,734
Fourth ,, ,, ,,	2,040

and concluded thus:- 'I am of opinion that paraffin oil, consumed in Doty's lamps, alike from readiness in trimming and lighting up, from steadiness of flame, and from high photogenic power, possesses decided advantages over colza oil for lighthouse illumination.'[10] In April of the same year, Dr. Macadam reported that he had got still more favourable results with the first and second order of lamps, by an alteration of the wicks. These experiments were simultaneous with practical trials of the oil at Girdleness, Pladda, and Pentland Skerries lighthouses. In December 1871 appeared a report by Professor Tyndall,[11] who had been employed by the Trinity House to examine into the relative merits of the lamp used by the Trinity House and that introduced by Captain Doty. Dr. Tyndall's experiments lasted fifteen days, and he thus concluded an elaborate report on the matter:- 'With oil No. 1, which appears to be more suitable for lighthouse purposes than any other, the Doty lamp shows a small but distinct superiority. As regards this oil, the simple changes introduced by Mr. Douglass into the Trinity burner appear to be almost as effective as the more elaborate devices of a central button and an external jacket introduced by Captain Doty. But as regards the combustion of oils richer in carbon, and, therefore, more liable to produce a smoky flame, the Doty lamp possesses, within certain limits, a distinct advantage. On the whole, therefore, I should pronounce the Doty lamp the most effective of the two.' Mr. Douglass' alterations of the old colza burner were as follows, according to Dr. Tyndall:- Altering the tips of the wick cases, introducing a perforated disc at the centre of the flame, and modifying the form of the glass chimney, also altering the level of the oil (an expedient previously adopted by Captain Doty). The issue of this report was followed by a voluminous correspondence, extending from 1871 to 1874, between the Trinity House, the Board of Trade, the Northern Commissioners, Messrs. D. and T. Stevenson, and Captain Doty, relating to certain claims on patent rights and proposals to settle the same.[12] In reference to this report, the question of vegetable oils *versus* mineral oil was considered by Dr. Tyndall to have been settled by previous reports of Dr. Macadam, Mr. Valentine, of the Royal College of Chemistry, Mr. Douglass, and Messrs. D. and T. Stevenson. After quoting the report of Mr. Douglass to the Trinity House, that with mineral oil used in a single-wick lamp an economy of 72.7 per cent, and with a four-wick burner of 60.5 per cent, had been gained, as against the results obtained by the use of colza oil, Dr. Tyndall remarked:- 'This consensus of evidence leaves no doubt upon the mind that, as regards cost and illuminating power, the paraffin light really possesses the advantages claimed for it.' Meanwhile, the French Government had been actively at work, and by the year 1872 sixty-nine lighthouse stations in France were being supplied with paraffin oil. In 1873 M. Reynaud reported to the Foreign Office[13] that Captain Doty had been awarded 10,000 francs by the French Government in recognition of his services; and in the same year the Duc de Broglie reported, through Lord Lyons,[14] that two kinds of burners were giving good results, viz., the Doty and a modified form of Fresnel lamp, made by M. Henry Lapaute; and that they cost the administration the same price, viz.:-

		Francs.
Five	meshes	60
Four	,,	50
Three	,,	40
Two	,,	32

In 1874 the Trinity House reported to the Board of Trade that paraffin oil was used in nineteen English lighthouses, and they recommended its introduction into twenty more, and a contract was entered into with Messrs. De Mille and Co. for burners, the prices being as follows[15]:-

		£.	s.
Five	wicks	10	15
Four	,,	8	0
Three	,,	6	0
Two	,,	4	10

In the same year, also, the Board of Trade sanctioned the Northern Commissioners introducing (through Messrs. D. and T. Stevenson) the Doty burner and mineral oil into all Scotch lighthouses. As regarded the economy of mineral oil for lighthouse purposes, its cost had always been under that of colza, or other fatty oils; frequently only one-quarter, and never more than one-half the price. In foreign stations, it would appear from the Paper that it was also economical, notwithstanding the expense of packages and freight. Mr. Brunton had remarked correctly, that much care was required in the selection of the oil, and in freeing it from the volatile constituents which were always present in crude or imperfectly refined mineral oils. If a standard of 145° to 150° flash point were too low, it would be easy to manufacture an oil having a flash point even as high as 250°, which was the quality of oil supplied for the use of saloon lamps of marine steamers. There was one point which the Author seemed to think of importance, viz., the heat produced during the combustion of mineral oil, as compared with that evolved during the combustion of vegetable oils. To use the words of the late Professor Rankine: 'The total heat of combustion of any compound of carbon and hydrogen is the same as the quantities of heat which the hydrogen and carbon contained in it would produce separately by their combination with oxygen, and when hydrogen and oxygen exist in the compound in proper proportion to form water, these constituents have no effect on the total heat of combustion.'

The average constitution of vegetable oils was:

Carbon	76
Hydrogen	12
Oxygen	12
	100

Applying the above rules, the available carbon and hydrogen for producing heat in say 100 lbs. of the oil would be:

	lbs.
Carbon	76
Hydrogen	10½

Again, the average composition of mineral oils was:

	lbs.
Carbon	85
Hydrogen	15
	100

Now the number of British heat units produced by burning 1 lb. of hydrogen was 62,032, and by burning 1 lb. of carbon 14,500, therefore the following calculation would show the total heat of combustion of vegetable *versus* mineral oils:-

Vegetable oils	Carbon	76 x 14,500 = 1,102,000
	Hydrogen	10½ x 62,032 = 651,336
		1,753,336
	Carbon	85 x 14,500 = 1,232,500
Mineral oils	Hydrogen	15 x 62,032 = 930,480
		2,162,980

Thus mineral oils evolved 23 per cent. more heat in burning than vegetable oils; but, on the other hand, taking into consideration the increased light they gave, the amount of heat evolved per unit of light was actually less with mineral oils than with vegetable oils, when the average results of various burners were taken as the basis of calculation. It appeared probable that the breakage of glass chimneys, observed by the Author, was owing to the fact that a mineral oil flame was generally more concentrated than that of a vegetable oil, thus localising the heat produced. He would suggest that if this difficulty caused too much trouble, the form of the chimney, or the construction of the burner, should be altered so as more effectually to spread the flame.

GEOLOGY OF JAPAN

At this meeting the PRESIDENT directed attention to an interesting series of 123 specimens of the Mineral Products of Japan, - a gift to the Institution from the Government of that Empire; whereupon it was

> Resolved unanimously, - That the best thanks of the Institution be recorded, and be tendered to the Japanese authorities, for the interesting and valuable Collection of Geological and Mineralogical Specimens which they had offered for the acceptance of the Society.

In speaking to this Resolution, Dr. POLE desired, as one of the English engineers whom the Japanese Government had honoured with their confidence, to remark upon the interest taken by the authorities in engineering works, and how actively those works were being introduced. It was a pledge of the earnestness with which this was being done, that the authorities were investigating very carefully the mineral resources of the country, as was abundantly proved by this admirable collection of specimens. A great deal of information had been acquired, which was not formerly possessed, showing that the country was exceedingly rich in mineral products. The Government had not confined their attention to small specimens for exhibition merely; for he had, within the last few days, received samples of limestone from Japan, which he had been commissioned to get analysed, with a view to ascertain whether hydraulic lime could not be obtained, so as to save the large expense involved in the importation of Portland cement from England. The specimens presented to the Institution would no doubt prove extremely valuable. Considering the connection of English engineers with Japan, which he hoped would long continue, it would be interesting and useful to know what the products of the country were, which could not be better shown than by the specimens presented. But there was another reason for feeling grateful to the Japanese authorities - the kindness shown by them to the large staff of English engineers in that country. Many members of the Institution, and other Englishmen and Europeans, were engaged there, and all gave but one account of the behaviour of the Government to them. Formerly their lives would have been in danger, but now they were perfectly safe, and were treated with the greatest courtesy and consideration.

His Excellency Mr. WOOYENO, The Japanese Minister, expressed through Mr. Nagasaki, his great gratification at the statement that the collection of minerals sent by his Government was valued by the Institution. He should always be glad to do anything in his power to promote the interests of so large and important body, and to advance scientific knowledge.

NOTES

* We have followed the spellings of the original edition throughout. ed.

1 These lights were erected by the Yokoska officials.
2 *Vide* Transactions of the Royal Scottish Society of Arts, vol. vii, p. 560.
3 *Vide* Transactions of the Royal Scottish Society of Arts, vol. vii, p. 557.
4 Report to the Imperial Government of Japan, relative to lighthouses by D. and T. Stevenson, Civil Engineers, Edinburgh, 21 June 1876.
5 *Ante*, p. 131.
6 *Vide* Transactions of the Royal Irish Academy, vol. xxi, p. 51; and Reports of the British Association for the Advancement of Science, 1847-1858, *passim.*
7 *Ante*, p. 116.
8 *Vide* vol. xxxix, p. 126.
9 *Vide* P. P. Correspondence relative to the substitution of mineral oils for colza oil in lighthouses, p. 7, 6 February 1872.
10 *Vide ibid.*, pp. 13 *et seq.*, 28 June 1871.
11 *Vide ibid.*, p. 23, 6 February 1872.
12 *Vide* P. P. Correspondence relative to the substitution of mineral oils for colza oil in lighthouses, *passim*, 27 June 1872; 6 February and 1 August 1873, and 22 May 1874.
13 *Vide ibid.*, p. 30, 1 August 1873.
14 *Vide ibid.*, p. 1, 22 May 1874.
15 *Vide* P. P. Correspondence relative to the substitution of mineral oils for colza oil in lighthouses, p. 130.

INDEX